immunity

The definitive guide to caring for your immune system

immunity. the science of staying well

DR JENNA MACCIOCHI

Thorsons
An imprint of HarperCollins*Publishers*
1 London Bridge Street
London SE1 9GF

www.harpercollins.co.uk

First published by HarperCollins*Publishers* 2020

3 5 7 9 10 8 6 4 2

A catalogue record of this book is
available from the British Library

ISBN 978-0-00-837026-8

Printed and bound in Great Britain by
CPI Group (UK) Ltd, Croydon

To Luca and Isabella who have given me so much in such a short time.

And to all those who believed in me when I struggled to believe in myself.

A Brief Note to Healthcare Professionals

In this book I have sought to present some of the current and big ideas in my field, fusing them with my own stories, personal anecdotes and my passion for this subject. My aim was also to convey a sense of wonderment for this silent system and the vast and complex science it encompasses. I am acutely aware that this is not – and could never be – an exhaustive text on the immune system. In my attempt to keep the narrative and scientific ideas as clear as possible for the general reader, I could not cover all the topics I wanted to or discuss them in the depth they may warrant. And rather than present a complete textbook-level description, I have deliberately chosen to focus on the broader contemporary issues – to bring an awareness to how modern lifestyles and our rapidly changing environment interact with and shape our immunity and, ergo, our health. In my quest to be evidence-based I have dived into the scientific literature, consulted with colleagues and reached out to other experts along the way, combining this information with case studies and alternative health literature to give a balanced and holistic overview. In some cases the only 'evidence' available is weak or at least not the most rigorous, so I have made efforts to state this where appropriate. Ultimately, any one book can only tell part of a story and this, I hope, is just part one.

Contents

CHAPTER 1

How to Be Well in a Modern World

'There is no permanent ideal of disease resistance, merely the shifting sands of impermanent obsolescence.'
MATT RIDLEY, author

Described by the US National Library of Medicine as 'the most complex system that the body has', our immune system is a vast constellation of cells and molecules spanning every nook and cranny of our bodies.

WHAT'S IMMUNITY AND WHY SHOULD YOU CARE?

'Immunity', from the Latin *immunis* meaning 'exempt', refers to the art and science of the human immune system – the various biological defence systems that keep us well. The immune system is a silent wonder. While we are very aware of our hearts beating and the breaths we take, we are much less aware of our immune system and the many jobs it performs to keep us well. This mighty system protects our health by resisting uninvited infections, maintaining order and balance in our bodies, and healing wounds. It is our foundation to wellbeing.

Immunology, the extraordinarily rich, yet complex study of the immune system, has changed the face of modern medicine over the last 30 years. There has never been a more fascinating time to delve into immunology and harness its power to improve your health. And that is what this book will do for you. Based on cutting-edge, evidence-based scientific research, this book takes you on an empowering journey through the latest astonishing discoveries in immunology. You will get to know the immune system inside out and come to an intimate understanding of how it works.

It is not often that one body system touches so many aspects of human biology in both sickness and health. Although it is seemingly esoteric and beyond comprehension at first blush, the immune system is the epitome of versatility and simplicity of purpose.

When it comes to maintaining good health, the immune system is our most precious asset, yet we rarely appreciate our essential defences until something goes wrong. We thank our immune system when we catch a cold and scramble for vitamin C supplements in the winter months. But what about all the other wonders it works? If it only existed for the first few weeks of chilly weather every year, we'd all be in serious trouble. Most of the time our immune system is working away in the background and we don't even notice it. Deeply entwined with every aspect of our physical and mental health, it lays the foundations of our health and longevity.

Immunity is a sophisticated system, with astonishing complexity. But there is a lot that can go wrong, from autoimmune diseases and allergies, to mental-health problems and metabolic issues – even cancer. We have an obsession with looking healthy and being well, and yet we are sicker and unhappier than ever before. That's because our delicately balanced immunity is easily compromised by our ever-accelerating pace of life, with its relentless stress, pollution, overconsuming and under-moving. What modern life deems 'healthy' may not actually be so. Today, we're more likely to die from a lifestyle-related disease than any other cause, and many of these would be preventable if we took better care of our health.

On our journey through modern immunology, you'll discover why the immune system is your sixth sense, connecting your health to your environment, feelings and emotions; why some people rarely get sick; what to do if you have a chronic illness; and what it really means to 'boost' your immunity.

My endless wonder at this vast and elegant system, integral to our wellbeing, has been the driving force behind both my career and personal life. From an early age I was intrigued by the human body, health and disease. My mum, a trained cook, strongly believed that health was in our own hands and taught me the old-fashioned cookery basics that have become a vital tool in my modern-day mum–life toolbox. She was the original purveyor of old wives' tales. Yet there was a wealth of untapped wisdom in her words that left me with a lingering curiosity, shaping my path as I embarked on my scientific career.

As I've discovered more about the immune system, I've found myself changing how I live in response to what I've learned. I'm driven to question what modern life tells us is healthy, yet makes our immunity go awry. Every day I find myself hunting for clues, researching the evolution of the immune system and thinking about the traditional ways of life that have shaped our health. Have we thrown the baby out with the bathwater by swapping old ways for new and shiny modern ones? This is a question I keep coming back to, and one which I'll answer in the chapters that follow, while also exploring how we can reclaim our innate knowledge of immunity through traditional rituals, supported by science. Keep in mind as you move through the chapters that the immune system is a bit like a wiggly moving octopus. Its undulating nature makes it difficult to cover it in a linear, page-by-page fashion. And one of its slippery charms is that the answer to many questions about it begin with 'It depends' or 'It's complicated.' Understanding the secrets of how to be well requires a balance between openness to new ideas (no matter how counterintuitive they may seem) and ruthless scrutiny of all concepts both old and new.

Despite its immense power, we often hear calls to 'boost' our immune system through food, exercise or a wellbeing practice. Although there remains much to learn about its interconnectedness and intricacies, to function well the whole system requires *balance*, not boosting. Read on as I explain what this means for your health.

INFECTION PROTECTION

The immune system is all that stands between us and the hordes of microbes (also known as germs, these are microscopic organisms including bacteria, fungi and viruses too small to be seen by the naked eye) that threaten us constantly. Hippocrates (460–377 BC) was the first to find a scientific explanation for disease. His was the theory of the four humours (yellow bile, black bile, phlegm and blood) in the body – the idea that these were in balance in healthy beings, but out of balance in a sick person. Sickness, he claimed, was the result of an excess of one of the humours, caused by *miasma* (poisonous vapours or mist filled with particles from decomposed matter) or *miasmata* (vapours coming from decomposing organic material such as waste, manure or dead bodies). This theory may sound strange to us now but reflects the limited tools and technologies available at the time. Without a way to 'see' germs, scientists didn't have much to go on. In the mid-19th century, as equipment for poking and probing evolved, 'germ theory' – devised by Louis Pasteur (the father of immunology) and his contemporaries – emerged and soon replaced the four humours in public consciousness. Germ theory is how most of us think of immunity: germs are bad, and immunity's white blood cells are our guardian protectors. But of course, nothing with health and immunity is ever that simple.

Immunity is hard-wired to detect and differentiate between what is us (self) and what is foreign (non-self), such as a germ. This makes it easy for our immune system to decide what to attack: non-self potentially dangerous germs that need eliminating, (normally) sparing our own precious tissues.

The immune system has had a complicated, you might even say fraught relationship with germs over the years. For centuries we have drawn a causal conclusion that microscopic germs cause illness – and for good reason. Devastating outbreaks, infection epidemics and perplexing illnesses have all been caused by the veritable smorgasbord of micro-organisms that we share this planet with. Over the last decades our fear of them has been realised, with devastating reports of swine flu, Zika, Ebola and many others – each outbreak prompting new concerns about infection protection. But from birth to death we are silently bombarded, minute by minute, with an untold number of potential infectious threats. Whether we get sick or not is decided by the integrity of our immune system. Most of the time, our immunity deals with these germs without us even knowing about it. That's how powerful the system really is. There are no medications that can protect us in our infectious world quite like our immune system.

THE IMMUNE SYSTEM: A REALLY, REALLY, REALLY BRIEF EXPLAINER

Though vastly complicated and mind-bogglingly confusing, as any of my immunology students will tell you, let's try and keep it real simple.

The immune system is not, in fact, one thing, and it's not in one place. Rather, the immune system is a whole galaxy of cells known as white blood cells (leukocytes, to give them their proper name). Immunity also includes lymphatic organs (such as lymph nodes, bone marrow and spleen), molecules (cytokines) and their collective array of biological functions. The bone marrow is your immune-cell factory where new immune cells are born from stem cells – a blank-canvas cell with the capacity to evolve into any one of your numerous immune-cell types. Despite their name, white blood cells are not just found in your blood, but in strategic locations all over your body. Every single one of them brings its own particular skill set, featuring an array of receptors and molecules capable of inducing the many different flavours of an immune response.

Border Control

A good way to start understanding the immune system is to liken it to a castle – a fortress with many layers of defence, all working as a close-knit team. Your body's borders act as the first line of immune defence: skin and the mucus membranes that line your natural openings – such as your mouth, nose and digestive tract – are part of your immunity, making and releasing substances that create a hostile environment for the invaders or attacking and destroying them directly. These physical barriers are as delicate as they are protective, with inherent vulnerabilities, and over the millennia just about every germ that wants to get in has evolved ways to help it do so. In response, our immunity – or castle – has devised a sophisticated system, choosing the best and discarding the least useful bits of our elegant defence mechanisms, so that every barrier surface has its own unique set of immune defences designed for that location.

Priority Setting: Full-Frontal Innate Immunity

Put simply, the cells and molecules of the immune system comprise two parts: innate and adaptive (which we will come to soon). Together, they are your body's safeguard. Innate response is what you notice when you first get sick (inflammation is an example of an innate response). It attacks with a vigorous, feet-first approach. What our innate immune response lacks in specificity, it makes up for in speed. The starting line-up features the innate cells. Like scattered 'listening posts', these are diverse and everywhere in our body. They act as sensors, hard-wired to detect anything out of the ordinary, eating germs and debris, calling for backup. Legions of these innate immune foot soldiers rush to assess a problem, attack it with potent killers to defend (and potentially harm) our bodies, causing significant collateral damage to our own tissues. Our immune system's beautiful destructive dance, with its characteristic heat, swelling, redness and pain, looks like the scene of a microscopic multi-car crash. As this happens, you may spot those familiar signs – a stuffy nose, sore throat, tummy ache, fever, fatigue or headache – and then go on to experience familiar symptoms such as increased mucus, pus and a cough.

Now, while inflammation is a vital immune response and fundamental to our health, it is, by design, only supposed to be an acute, short-term assault. This is because it is as damaging to our own tissues as it is to invaders. Triggered incongruously, it can cause problems long after the initial danger is gone. As we'll explore, inflammation is core to our modern-day health niggles, existing on a continuum: from mild symptoms such as weight gain and fatigue at one end, to heart disease, depression and autoimmune conditions at the other.

The Path to Resolution

When we are sick, the body needs inflammation. But if this response continues for too long, it becomes counterproductive. The imprecise inflammatory defences discharged by innate cells can wreak havoc on our delicate tissues and even on entire organs. The immune system needs at some point to reduce the inflammation and has therefore acquired the tools to regulate it. So how does inflammation lessen? Let's explore.

Acute inflammation and all its associated chemical weaponry also upregulates many anti-inflammatory mechanisms to temper the fire. Special pro-resolving mediators (signalling molecules) are naturally produced in the body during the process of inflammation. And they, ultimately, close the loop on the whole process. Rather than turning the immune system off (and risking infection), resolution works in collaboration with inflammation, letting it do its job before gently inviting back the status quo.

Common over-the-counter anti-inflammatories such as ibuprofen or acetaminophen (paracetamol) block important pro-resolving signals, inhibiting the way our bodies naturally resolve inflammation. This is why they are not recommended as a long-term treatment. Aspirin, on the other hand, does not block but gently attenuates, while also stimulating resolution – which is why it is sometimes recommended at a low dose for certain inflammatory conditions. Resolution of inflammation might not seem like such a big deal, but it's paradigm-shifting, with vast potential to aid our modern-day health crisis, controlling chronic inflammatory disorders and promoting wound healing. We can support the resolution of inflammation in other ways too, which, as you will read, may be our best chance during future health challenges.

Slow and Specific Wins the Race

Innate immunity is pretty darned good at quickly detecting and eliminating potentially harmful germs, sensing and repairing damage, and quietly removing old or malfunctioning cells. But although it provides immediate protection against intruders, it is incomplete at best. When stressed to the limit, it has to call on its big brother: adaptive immunity.

If innate immunity is an initial shotgun blast at all the bad guys, adaptive immunity is a targeted missile. It's a second-line defence that takes a while to kick in – some five to seven days. Unlike innate immunity, which can involve a veritable horde of white blood cells, adaptive immunity is controlled by the lymphocytes. There are two categories of lymphocytes, known as B lymphocytes and T lymphocytes. T lymphocytes are the master controllers, sent out into the body, controlling the levers of many other arms of immunity; B lymphocytes are our special reconnaissance unit – the antibody producers.

Each of us has a unique repertoire of T and B lymphocytes, adding a layer of individual exclusivity to our immunity and, consequently, our health. An immune system that doesn't produce a huge variety of unique T and B lymphocytes will probably miss or 'not see' certain germs or viruses, and these could go on, unchallenged, to cause disease. This happens during ageing, for example as we will explore in the next chapter. Adaptive immunity is a remarkably effective process. But much like the innate inflammation response, when it is improperly triggered, it's equally bad for our health, as we'll see.

Blurring the Lines

To fight off infection and disease, the innate and adaptive parts of our immune system collaborate to detect and destroy anything recognised as foreign or dangerous in the body. We first need to co-opt the innate immune system to capture germs and present them via some oh-so-important compatibility molecules (more on these in a bit) to T lymphocytes waiting in anticipation. The innate immune system is an ingenious doorbell that awakens the adaptive immune response with germs in a format they can recognise.

T lymphocytes then swing into action, cloning themselves into huge armies to maximise their fighting power and differentiate into specialised subsets – driving, fine-tuning and sometimes regulating other immune cells – to maximise the chance that intruders are eliminated at minimal cost. Lymph nodes bring B and T lymphocytes together with innate cells at the right place and time. This is why your lymph nodes become swollen when you've got a cold coming on.

Collectively, the immune system is, in fact, a model of versatility rather than rigid divisions. Feel those swollen lymph nodes next time you have a sore throat and congratulate yourself on successfully linking the innate and adaptive immune systems.

How Is Your Memory?

Innate immunity is vital for life, but it only has a short-term memory. So instead of mounting a faster and more effective response upon encountering a known trespasser, it starts sluggishly from scratch each time. The adaptive immune system, however, is like a library of 'memory' cells. Information on every virus, bacteria or fungi that ever invaded your body and was defeated by your immunity is archived by this immune system, identifying them by their molecular shape. This is called immunological memory. And it almost never forgets!

Memory cells do not actively fight the current infection. They patrol the body in case of future infection with the same germ, so when we face that same disease-causing germ, this retained memory means we know how to defeat it – often even before we've experienced any symptoms. This explains why you only catch certain diseases once, such as (in most cases) chickenpox, despite repeated exposure. More cunning viruses, however, like influenza and rhinovirus (the cause of the common cold), have developed sneaky ways to evade our immunological memory by continually changing their molecular information.

Natural Born Killers

Natural killer (NK) cells are, as their name suggests, killing machines. Unlike T and B cells, these are not produced in response to a specific antigen (toxin). Instead, they recognise changes in our own cells. These cells are crucial to our health, rapidly responding to kill our own cells when they get infected with a virus.

NK cells are our bodies' main cancer-surveillance tool. With specialised receptors, they patrol the body inspecting each cell. They respond to newly formed tumours and abnormal growth of our own cells. And they also decide the fate of a pregnancy. NK cells represent 10 per cent of white blood cells but are also found dotted around the body in various places including the liver, lungs and lymph nodes – like your very own Special Forces regiment, with a smaller number of troops but just as deadly as an entire infantry division.

Really Tolerant T Cells

Military metaphors can be a useful way to describe our immune defences, but only when we use them in a nuanced way. The immune system is more sophisticated than a ruthless defender using brute force. It is, perhaps, more of a peacekeeping force, seeking a steady, harmonious state, not a constant battleground. Although it focuses on tossing out the bad guys, it must do this while causing as little damage as possible. This is to avoid damaging our precious tissues.

So who guards the guards? At the top of the peacekeeping chain are T regulatory cells, known as Tregs. Tregs control or suppress other cells in the immune system, fighting substantial fires and making sure the other cells toe the line. Tregs are designed to send a signal that immunity should withdraw, pause an attack and stand down.

Without these regulatory cells and molecules, the state of inflammation that helps destroy threats would lay our bodies to waste. Tregs are crucial to the overall balance of our immunity. They maintain tolerance to 'good' germs, prevent autoimmune diseases and limit inflammatory ones. But they can also suppress the immune system, preventing it from doing its job against certain infections and tumours. For optimal health, we require sufficient but not excessive regulation by Tregs. Imagine a small proportion of your immune system is reserved for regulation. Some of that is genetically determined, but some can be shaped by life choices like diet and exercise, stress and sleep. The system therefore operates on a constant tightrope: insufficient immunity (too many Tregs) may increase incidence of infections and cancer, while excessive 'friendly fire' (too few Tregs) may lead to damage to our own cells and organs.

Vaccination – The Great Teacher of Community Immunity

Vaccination is one of public health's greatest success stories. Yet it also has become one of the most contentious areas of health policy over the past two decades and few things are more divisive in immunity than vaccinations (see Chapter 2 for more on this).

Vaccines work on the same principle as acquiring an infection naturally. When you acquire an infection naturally, your adaptive immunity, T cells, B cells and antibodies, are primed to fight off the germ. Then, a residual immunological memory remains, patrolling your body for decades (perhaps even a lifetime), ready to protect you if that germ ever tries to infect again. Your immune system reacts to the vaccine in a similar way to being invaded by the disease itself. Introducing a bit of weakened virus or bacteria teaches the immune system how to deal with it. It generates a memory response so when the real germ comes along, an army is at the ready. Just as each germ is unique, each vaccine is too. The memory produced by vaccines is different for each type. This is why some require booster shots – because vaccines don't 100 per cent replicate the natural course of an infection, they (mostly, depending on the vaccine) don't all provide the same degree of immunological memory.

Why Do Some People Never Seem to Get Sick?

In my job I like to practise what I preach. Most of the time I pretty much have my health nailed down, but each year I still catch the odd cold or bug. The average adult will get two to four colds a year, but some lucky people claim never to catch a cold or take a day off sick. We've all met them, whether it's through family or at work – they breeze through cold and flu season without so much as a sniffle. What is their secret? How can we be more like them?

Your immunity is unique to you and determined by several factors. Genetics plays a part, but not in the way you might think. We each have around 25,000 genes, but your genetic code only differs by about 1 per cent from that of the person next to you. You might think that the majority of these differences are related to why we all look different, varying our hair colour, height or personality. But, aside from our brains, the genes that vary most between us are, in fact, a tiny cluster that plays a disproportionately large part in our health. This gene cluster is called the human leukocyte antigen (HLA), also known as the major histocompatibility complex (MHC). Let's call these genes the 'compatibility' genes for short.

Our compatibility genes encode our immunity, yet they are as changeable as the pathogens (bacteria or viruses) they aim to prevent. These molecules have evolved to come in a variety of shapes and sizes. They mutate with each generation, unlike any other genes in our bodies, in an effort to track the ever-changing infectious threats we face. Our immunity depends on their combined performance. Our compatibility genes flag viruses and bacteria, and present them to the immune system for eradication. The fact they come in a range of shapes and sizes, prevents a virus or bacteria from mutating and then evading the immune system in every person they infect.

These special genes tell us a lot about the intricate balancing act essential for not only our health, but the survival of our species. Humans are incredibly similar but also fundamentally diverse, and compatibility genes are the key to our individuality. Simply put, if all our immune defence systems were identical, a single deadly disease could come along and wipe us all out. But this ingenious process comes with certain trade-offs. For example, we can't easily exchange our body parts; if you have ever received an organ transplant, you will be familiar with the life-long challenges involved in stopping your body rejecting the new tissue. In fact, immunity that works wonders for protecting one person can be quite deadly in the life of another.

Compatibility genes are the reason why we all fare differently when faced with the same infection. You may have inherited a set of compatibility genes that are expert at dealing with one particular virus – the common cold, say. This does not mean that your immune system is better or worse than mine; just that yours would deal better with fighting the seasonal lurgy. But when it comes to a different type of germ, I may fare better. And it's not just our individual susceptibility to infection that is affected by our own unique combination of compatibility genes. For example, there are genetic variants in immune compatibility genes which protect some people from the HIV virus but leave them with an 80 per cent chance of getting a horrible autoimmune disease called ankylosing spondylitis.

Like our fingerprints, immunity makes us truly individual. The inherent diversity in how our immune systems respond to different diseases is an entirely deliberate design by mother nature without hierarchy. While the world is busy arguing over the physical, visible differences between people, our immune compatibility genes don't discriminate. No one has a better or worse set. It's the collective diversity that's crucial. This diversity has been carefully crafted by evolution. Millions of years of evolution have more or less tuned our immune system to be running optimally. We might not be able to become completely resistant to disease, but it's why we haven't all been wiped out already.

Our immunity has been in place for around 500 million years. Its history goes so far back that we share it with other jawed vertebrates. And it has remained largely the same ever since it first evolved. Our immunity is millennia old, shaped and polished by evolution, and, by virtue of this, it is very good at what it does. The fact that our immune system is largely untouched by evolution is a testament to how important and effective it is and always has been. Evolution is not a predetermined design process; rather, it is one that advances in fits of trial and error as well as chance and necessity. Your successors will carry in them not only the final product of these efforts, a one-and-only perfect immune system, but also the marks and remnants of many immune systems.

If the Immune System Protects Us, Why Do We Still Get Sick?

Most of the time, our bodies win the fight against germs, but sometimes they lose. People who 'never get sick' may catch a slight cold now and again or suffer an occasional ache and pain. This makes sense when you realise that we live in a microbial world, and the microbes were here first.

Microbes haven't only threatened our health, they have shaped it. They were the first traces of life to appear. And much later on, from within those early microbial ecosystems (and never separate from them), larger multicellular organisms evolved. Including us. And we have never quite separated from them.

Now we know that there are up to 1 trillion microbial species on Earth, and only a minute fraction of them cause disease. So vilifying all microbes for the sake of a few is a mistake, and possibly one of our biggest (you'll discover why in Chapter 3). Nevertheless, much of the fear of microbes lingers in the public consciousness. As I mentioned, it's quite normal to get up to four minor infections (like a cold) each year. But as infections are beaten back by modern-day sanitation, vaccination and antibiotics, we now have huge increases in 'non-infectious', lifestyle-related diseases. As you will read, this is no coincidence.

Let's look at how infection spreads. Take rhinovirus, the cause of the common cold, for instance. Roughly one in five people carries the rhinovirus at any given time in the tissues of their nasal passages (the prefix 'rhin' in Greek literally means 'the nose'). To infect you, these germs need three things:

- A way to get out of the reservoir (the person sitting near you who is sick)
- A mode of transportation to a new home (that person sneezing – a sneeze produces 40,000 droplets and you can get infected by inhaling just one)
- A way to get into their new home (you)

Another classic route for the spread of germs is poor hygiene, especially inadequate hand-washing. Bacteria are passed to everything we touch. But simply being a thorough hand-washer with good personal hygiene helps you stay healthy and avoid illness-causing bacteria. And while we can't account for how infected people behave around us and we can't control for the immunity genes that we have inherited, we do have some control over the various lifestyle levers we can pull to get the best out of our defences.

Nature Versus Nurture

Like fingerprints, immune systems vary from person to person. We all inherit a unique set of immunity genes – but this is just an instruction manual, and while the code can't be changed it can be interpreted in many ways. You can help train and maintain it. Epigenetics (the study of how our genetic code is interpreted) is manipulated by many lifestyle factors. An example of epigenetics might be the changes in something called DNA methylation. (Essentially, methylation is like handcuffs for a gene, marking that gene as not to be used. If certain genes are methylated, then a cell might not be able to turn on critical functions and could become a cancer cell, for instance.) Various environmental factors can alter our methylation pattern, such as smoking, poor diet, air pollutants and alcohol. Adjustments in our epigenetic patterning can derail some of our vital immune responses. The metaphor used by scientists is: 'Your genes load the gun, but environment pulls the trigger.' Just a little immunity TLC from time to time can, for most people, keep all components functioning more smoothly.

Despite the prominent role played by genetics, immunity is not fixed by our genes. It's continuously nurtured by our encounters and adventures, shaped by our changing emotions and surroundings, responding to how we live our lives.[1] As mentioned, our immunity even has the capacity to learn and develop a memory. Collectively, this plethora of influences accrued across a lifetime determines, at any given point, if we get sick and for how long. This human 'exposome' is our immunobiography (more of which later), the culmination of everything that challenges our health throughout our lives, including infection, our diets, lifestyle factors and social influences. In short, it's what we refer to as *nurture* (nature being the genetic part). This holistic view of health and disease goes some way to explain why some people end up being more susceptible to getting sick.

INFECTION – THE FACTS AND THE FICTION

It starts with a sniffle. Next thing you know, the whole household is sneezing, coughing and passing tissues. Before long, the common cold, or worse, the seasonal flu, is upon you all. Wondering how you can dodge those germs or 'boost' your immunity? With so much conflicting advice out there, it can be difficult to know what to believe. So let's start by putting myths to rest with a simple guide.

No. 1 Tip to Avoid Infection

In the mid-19th century, a Hungarian man called Dr Ignaz Semmelweis made an important discovery. He linked childbed fever among women who had just delivered their babies in the hospital to the failure of the doctors who delivered those babies to wash their hands after performing anatomical dissections of the dead. The key to reducing infection was simple: wash your hands. Semmelweis died before his theory gained support, but hand-washing remains a cornerstone of modern-day infection control.

Most People are Infectious Before They Get Symptoms

Before the onset of symptoms, a person may already have an infection. How this works depends on the germ in question. Respiratory viruses are best engineered to spread when you have physical symptoms; the more symptomatic you are, the more you sneeze and cough, the more likely you are to spread an infection. But sometimes you can infect others with a virus even when you are asymptomatic. For influenza, you are infectious one day before symptoms and five to seven days after their onset. Young children and patients with altered immune systems can spread the virus for longer periods of time.

It's Just a Mild Cold – Am I Less Infectious?

Just because you have mild symptoms doesn't mean the virus is mild. It may just mean that you lucked out with your compatibility genes for that particular germ and your immune system is able to control the infection. For this reason, it's important to remember that even with minimal symptoms, you can still infect others and potentially make some people really sick, especially those who are vulnerable, including the elderly and very young.

Why Do Colds and Flu Strike in Winter?

The flu season in the UK starts in October and is in full swing by December, generally reaching its peak in February and ending in March. For our Antipodean cousins, the season is flipped. Put simply, whenever and wherever there is winter, there is flu. Even the word 'influenza' may be a clue – its original Italian name, '*influenza di freddo*', means 'influence of the cold'.

A common misconception, however, is that flu is *caused* by cold temperatures. This is not quite true. Flu is *caused* by the influenza virus. Cold temperatures simply create conditions that make it easier for the virus to spread. Scientists now know that the influenza virus is transmitted best at cooler temperatures and low humidity. So the viruses that cause the flu survive better in winter and are able to infect more people. Another reason is the lack of sunlight and the different lifestyles we lead in winter months: days are colder and shorter during the winter, so we spend more time indoors and are more likely to share air with someone who's infected. Lack of sunlight leads to low levels of vitamin D, a key immune-nourishing nutrient. Having less vitamin D decreases our immunity's ability to fight the virus.[2]

You'll Catch a Cold from Being Cold

'Put on a jacket or you'll catch a cold.' I'm sure you were told this as a child, but nowadays it is usually dismissed as an age-old misconception. Much like the previous point, it's the virus, not the temperature, that makes us sick, right? Well, yes. But it turns out that this advice contains a kernel of truth too. It has been shown that when you are exposed to the cold for a prolonged time you may not be able to launch the most robust immune attack. While this effect might be marginal in most healthy people, older people, young kids or those with underlying health issues may have an even harder time fighting off the virus. So, best heed your parents' advice and bundle up when heading out in the cold! Wearing a scarf in winter does warm the air in the back of your throat, making it less hospitable for those seasonal viruses that prefer cool air.

Disease Symptoms Are Sometimes the Immune System Doing Its Job

You might have heard that bacteria, viruses and fungi are the cause of disease symptoms, but often this is technically incorrect. Disease symptoms mostly occur because your immune system is at work reacting to trespassing micro-organisms with inflammation.

Take the example of the common cold. Your immune system jumps into action when the cold virus (rhinovirus) invades the epithelial cells that line your airways. Immune system chemicals dilate your blood vessels and increase their permeability, allowing proteins and white blood cells to reach the infected tissues in your nose, sinuses and throat. However, this causes nasal congestion. Additionally, you may get a runny nose because of the increased fluid leakage from your permeable capillaries, combined with increased mucus production triggered by the immune response itself. Then there is fever.

Through evolution, we developed fever as a response to infection. Raising your temperature makes most germs less efficient at multiplying and helps your immune cells work more effectively. There is evidence too that fever even improves the ability of antibiotics to kill bacteria.[3] Awareness of this phenomenon has given rise to two appealing, but conflicting, schools of thought. One theory is that fever is a natural part of the immune response to infection, so we shouldn't interfere with it. The other is that fever is a potentially harmful consequence of infection, so we should suppress it to minimise any complications.

What does science have to say about these conflicting theories? There is actually little evidence that fever itself, even a high one, is harmful.[4] Unless you are really uncomfortable, have difficulty breathing or have a fever lasting more than a few days it's normally wise to avoid fever-reducing medications like paracetamol or ibuprofen as they are not without side effects. The immune system can still get its job done if you take medicine to reduce fever, but slightly less effectively – and taking it won't speed up the process.

Over the Counter

No one wants to stay at home feeling miserable with a cold or the flu. And sometimes you just can't. So what about all those over-the-counter (OTC) cold and flu medications? Most OTC cough syrups are not effective for adults or children,[5] and the NHS now recommends honey rather than cough syrup. OTC cold and flu medications won't 'cure' or even shorten your cold or flu, but there is some evidence that they may provide some symptom relief in adults. Each time my husband catches a cold, he treks to the pharmacy, spending (wasting) lots of money searching out a cure. Perhaps the idea that he is proactively looking after himself makes him feel better about the situation, which may create a little placebo effect. But the pills he buys won't cure his cold.

Reducing symptoms is, however, often desirable, or necessary if life demands that you function. So if work really can't wait, which product is best for your symptoms? Here's the lowdown:

Antibiotics definitely don't help and shouldn't be used for the viruses that cause colds and flu. Cold and flu tablets may contain decongestants, pain relievers, antihistamines and cough suppressants. So think carefully about your main symptoms when selecting a product and read the label carefully – avoiding taking medicine you don't need reduces the risk of side effects. Decongestant tablets can have effects beyond the nose and may exacerbate other medical conditions such as high blood pressure. Speak to your pharmacist before taking these medicines.

When it comes to immunity there are few quick fixes. The minor benefits from certain treatments need to be weighed up against the risk of side effects and the cost of the medicine. There are a few nourishing ways to ease a cold without pharmaceuticals but rest really is the best medicine. If only you can step off the treadmill of life for long enough.

DO YOU NEED AN IMMUNE 'BOOST'?

If a well-functioning immune system is key to living a longer and healthier life, then the idea that you can boost your immune system is certainly enticing. Searching 'immunity' online or walking through the aisles of any health-food store reveals an abundance of nutritional supplements, cold remedies and fortified foods all promising to 'boost' your immune system and, by doing so, to stave off colds and flu. But is there any scientific truth in these claims?

Sadly for consumers of these 'immune-boosting' products, the notion of the immune system as some kind of internal force field that can be easily ramped up makes little sense scientifically and is one of the biggest misconceptions about immunity that I come across in my work.

Because of the way your immune system is designed to work, you definitely wouldn't want it to be boosted. Taken literally, the term 'immune-boosting' is an unfortunate one. Rather than a single binary on/off switch, the immune system is more like a series of switches – a rheostat requiring constant tweaking to get it just right. Attempting to 'boost' the cells of your immune system is complicated. There are so many different kinds that respond to so many different microbes in so many ways. Which should you boost? And how much? So far, there is no single answer.

Researchers have looked at whether things like echinacea, green tea, garlic and wheatgrass supplements can help us see off germs, but existing evidence in support of a single immune-boosting nutrient or superfood is not strong. If you are looking to strengthen your immunity, the best way is through a combined approach based on the information in the following chapters'.

NOT ALL GERMS CAUSE INFECTION AND NOT ALL DISEASE IS FROM GERMS

Not all germs are bad and not all disease is caused by germs. Although germ theory (see p. 4) still dominates how we think about immunity, we know that things are more complicated than that.

The Danger Theory of Disease

It would be nice to have a simple, clean separation: immunity doesn't react to our own harmless 'self' molecules, but only to things that are not us – i.e. harmful 'non-self' germs. But is this the case? Are all harmful, non-self, immune-triggering molecules from infectious germs? To complicate matters further, not all diseases are caused by foreign germs (take allergies, for example). And not all germs in the body warrant destruction (the microbiota say, that you will meet in Chapter 3). Various aspects of our day-to-day lives involve interaction with something non-self – and many are things that we don't want our immune systems to reject: mothers don't reject foetuses; we tend not to attack the food we eat; and we don't reject the millions of microbes living harmlessly, and often helpfully, on and in us. All these are examples of foreign 'non-self' things that somehow evade the immune system's wrath. Yet non-infectious immune diseases related to modern lifestyles are rampant. Until relatively recently the field of immunology was agnostic as to how this happened. Then, along came Polly.

While for decades we wrongly assumed that the immune system could simply spot self and differentiate that from non-self, a 20th-century immunologist rewrote the script. Born in 1947, Polly Matzinger blazed an unconventional trail. Following stints as a bass player, a dog trainer, a carpenter, a biology student – and later a Playboy bunny – she went on to change our understanding of how disease happens, poking holes in the dogma and becoming one of the most influential thinkers in the field. It turns out that immunity is not just concerned with whether the thing it is reacting to is self or non-self; rather, it is also hard-wired to detect patterns of danger.[6] Why is danger relevant? We can get sick in the absence of an active 'non-self' infection. Which means that something is still triggering the immune system when germs are absent. This trigger, Polly discovered, is a pattern of danger signals from our own cells and tissues.

In fact, we even have a term for this: danger-associated molecular patterns (DAMPs).[7] Danger comes from any change in our cells and tissues, such as damage or stress, and can even be from lifestyle-associated molecular patterns (LAMPs).[8] The immune system's smoke detectors sense something is awry. This triggers a series of molecular signals which bring about an immune response. One example of how this happens is when inflammation occurs in the absence of any germs.[9] We call this sterile inflammation. You might have experienced this, as I did recently after tearing the ligaments in my ankle. No broken skin and no infection, but immediate swelling and visible inflammation as my painful ankle ballooned and bruised. The missing 'danger' link in this situation is in part due to the mitochondria in our cells. The sheer mechanical tearing up of tissues (like when I twisted my ankle) releases the contents of our cells into the blood, including their cellular battery packs, the mitochondria, which are distant cousins of bacteria. In terms of our immunity, this means they are molecularly 'non-self', despite being part of us. Why is this important? Well, inside our cells, mitochondria are shielded from our immunity, but when they leak out, they are recognised (by innate immunity) as patterns of danger. Their bacteria ancestry makes them exceptionally good at signalling to start inflammation, even though no infectious germs are present.

Sixth Sense: Immunity Beyond Infection Protection

Understanding infection protection was the first step in discovering how the immune system works, and it remains the main lens through which our immunity is viewed. But infection protection is only part of the story. Vigilantly patrolling, our health guardian is always on. Beyond just surveying for possible infections, immunity has many additional vital functions: it governs repair and healing, regulates body weight and metabolism, defines how we age and if, or when, we might encounter age-related disease. Immunity also determines the success of a pregnancy or a transplanted organ. And, crucially, using those NK immune cells you met earlier, it is our main cancer surveillance system. The immune system is so fundamental to our existence, it's the essence of our species.

People often talk about having a 'sixth sense', a way of perceiving the world other than by the standard channels of seeing, smelling, tasting, hearing or touching. I'd like to propose that your body has such a sense: the humble and hard-working immune system deserves recognition as a sensory organ of sorts. Much like vision, hearing and touch, our immune system recognises challenges from our environment and signals from our emotions. Unlike other types of cells, immune cells can sense and respond to environmental, hormonal, nutrient and even brain signals. This dynamic ability makes the immune system both fairly unique and quite peculiar.

In fact, science has started to consider immunity as a kind of second brain: a specialised network of biosensors designed to pick up information from within and around the body and relay it to the brain, where it can motivate and manipulate us to behave in specific ways. This may be as simple as huddling, shivering or sleeping – actions that are typically considered general symptoms of illness or debilitation, but may, in fact, be highly specific and beneficial responses to infection. In other cases, the immune system may inspire behaviours not typically associated with illness at all, such as passivity, aggression or sexual attraction. The immune system, it would appear, has quite a bit to say not only about how we behave but about how we feel and even who we are.

Instead of recognising light, sound waves or taste, the immune system recognises molecular shapes of germs, damage or danger – perturbations to our status quo. Sampling the environment, immunity has evolved rigorous mechanisms enmeshed with all our other body signals to keep us well. No other system in the body places such a premium on adaptability (except, possibly, the brain – and why would nature disconnect two such vital systems from one another?). If our immune systems can't counter the diversity in our ever-changing environment by adapting and responding, we're dead ducks.

Although it may sound counterintuitive, immunity keeps you alive, but is also disease-causing. Tip it one way, and your immune system weakens, leaving you vulnerable to pathogens. Tip it the other way, and it becomes overactive, attacking your own healthy tissue or benign things in the environment. Or it becomes unable to regulate the daily inflammatory wear and tear. These are not things you want.

Modern Malaise

Illness is part of our natural history – a biological heritage, crafted by evolution, the greatest part of it prior to the antibiotic era. Pre-antibiotics, infectious disease had the potential to wipe out the majority of us. This created evolutionary pressure to mount strong immune responses. But now, mostly everyone in the Western world survives the childhood infections of times gone by. As we humans have evolved, so too have our diseases. Over the past hundred or so years, deaths related to infectious disease have been dramatically reduced. But the diseases that have challenged us have also shaped us and left us with deeply embedded evolutionary vulnerabilities. And while epidemics of diphtheria, plague, cholera and smallpox no longer sweep across the Western world, we now have other foes to fight.

Today, germs are no longer the enemy. In fact, 99 per cent of those that surround us at any given moment (and they are everywhere, all the time) do not cause disease. Rather, they are 'good' bacteria – our microscopic health allies. This pokes holes in the 'all-germs-are-bad-and-should-be-avoided' dogma. These microbes were here first and our long-shared history with germs has cultivated an enduring marriage of mutual exploitation (we'll explore this further in Chapter 3). The spectrum of disease faced by our immunity today has changed considerably from the infectious ones that challenged our health in the past. Germs may no longer be the main enemy we face but the war on our health certainly isn't over. We have merely delayed and mostly transformed it.

Of all the major health threats to emerge, none has tested the foundations of public health so profoundly as the rise of chronic non-communicable diseases (NCDs), which include autoimmunity, allergy, metabolic dysfunction and some cancers. Immunity has a central role to play across the spectrum of these chronic conditions that have quickly replaced infectious diseases as the leading cause of premature death and disability worldwide. We are getting sick in the absence of infection – the Western world is experiencing a rising tide of lifestyle-related illnesses, with immune-triggering danger signals rife and immunity going awry. This trend is deadly, carries crippling economic costs and cannot be easily reversed under the conditions of our 21st-century lifestyle. While an infectious disease can spread fast and far and kill irrespective of age, complex modern NCDs are long-term conditions that develop silently over years – even decades in some cases – ending the lives of people still in their prime, figuratively if not actually killing them. But we should remember that for all the ways in which we seem to be sliding backward, we're lurching forward too. Since the 1980s, in Western countries we have sharply cut down on deaths from infections and malnutrition. Thanks to medical advances, rising wealth and much else, infectious diseases now kill fewer people than non-communicable ones. Sounds like great news, right? Well, it would be. But we have simply replaced one with the other, to the detriment of our health.

The good news? Doctors, scientists and the general public are increasingly aware that this is happening. Getting a handle on your health earlier in life might just be the key to better outcomes for these long-term chronic conditions. But although these modern-day health challenges are linked heavily to lifestyle, we cannot ignore how life in the 21st century also makes it difficult for people to change what they do. We can ban smoking, clean the air, build cycle lanes, tell people to eat '5-a-day', but it's not always easy for people to alter their behaviour, and this is a big part of foiling non-communicable diseases. It's also one reason why these diseases don't always generate the same level of concern: they're often regarded as the sufferer's fault.

Autoimmunity: Friendly Fire and a Case of Mistaken Identity

You may have heard of autoimmune diseases – but what exactly are they? And who gets them?

Autoimmunity is like friendly fire – a condition in which the immune system is too active, and unchecked T and B cells grow overzealous, wreaking havoc on the body they're supposed to protect. The body has a variety of fail-safes to ensure that the immune system doesn't attack itself, the guardian Treg cells being a mainstay. But sometimes these safeguards don't work or get weakened and we get an unwarranted inflammatory response towards ourselves.

The incidence of autoimmune disease is increasing worldwide by up to 7 per cent every year. Autoimmune disease is a large umbrella term, and there are over 80 diseases within it that are linked in some way. Each of these can be terrible, frustrating, debilitating and hard to diagnose. They are driven by the adaptive immune system with its long-lived memory, making them impossible to cure. Some of the most common ones include type 1 diabetes, rheumatoid arthritis and lupus.

The causes behind autoimmunity remain broad and murky, but what we know for sure is that there is rarely a single cause.

In fact, some research suggests that all of us have a little autoimmune potential as a trade-off for having over 100,000 unique T and B cells, each with the potential to detect a different molecule. This gives us a broad-spectrum infection protection but at a potential cost, in that some of these white blood cells might mistakenly recognise molecules on our own cells. Most of the time our specialised immune regulatory systems act as a firewall that keeps any rogue cells with autoimmune potential in check. Any issues with immune regulation leave us wide open to any counterintuitive attack. A big clue that this is the case lies in the fact that people who develop one autoimmune disease are more likely to develop another, indicating a global immune dysregulation is at play throughout the body.

Genetic predisposition accounts for roughly 30 per cent of all autoimmune diseases and currently there are more than 20 associated genes, many of which are in the 'compatibility' genes (see p. 14). This goes a long way towards explaining why autoimmune diseases tend to run in families. One family member might have rheumatoid arthritis, one might have thyroid disease and one might have lupus. It might be the same gene causing each of these different autoimmune conditions. But then identical twins are identical in genes, but they are not identical in developing autoimmune disease. This tells us that genetics is just part of the picture.

The remaining 70 per cent of autoimmune diseases are the result of environmental triggers and other situations we encounter – for example, toxic chemicals, diet, infections, digestive issues, stress and other lifestyle factors. Exposure to the perfect storm of non-genetic factors may actually flip the switch to push the immune system in the direction of an autoimmune condition. These factors are referred to as the 'mosaic of autoimmunity'. And to complicate things further, each of us is different. So the particular combination of factors that causes one person to develop an autoimmune condition may not do so in another. In other words, there's no one recipe for autoimmunity.

In the following chapters I will explore some of the ways that we can reduce our risk for autoimmunity. None of these lifestyle modifications is a guarantee to prevent or stop autoimmune disease. But together they may help to lower your risk.

A big issue with treating autoimmunity is that the slow-burning symptoms take so long to become apparent, and by the time they do, your immunity has silently been damaging your tissue for years. If you are unfortunate enough to be diagnosed with an autoimmune disease, you'll probably be told that it's incurable but manageable. Many of the current therapies for treating autoimmune diseases work by suppressing the entire immune system and are likened to using 'a cannon to swat a fly'. Consequently, while these therapies can successfully block an inflammatory response, they have the unwelcome side effect of leaving patients vulnerable to infection. Though there are many essential conventional therapies now available, I'll share some complementary approaches to treatment throughout the book.

Allergy – The Scourge of Modern Life

Allergies are the close cousins of autoimmunity. They are the result of the immune system overreacting – not to us, but to benign things in our environment like pollen (in the case of hay fever) or to a specific food (such as peanuts). Allergies are immunity overshooting, creating inflammation in the body in response to something that doesn't really pose much of a threat.

Our ancestors didn't suffer from allergies, and anyone from Generation X will recognise that they were extremely rare even a few decades ago. In fact, before the 1990s, peanut allergies were so rare that there was barely any data on them.

Incidence of allergies is now steeply climbing, showing no signs of abating. Like autoimmunity, this happens partly through genetics. Our collective population genes have changed little over this relatively short period of time, so can't be to blame. It is likely that these modern-day epidemics are due to an environment and lifestyle that do not allow our immune systems to properly calibrate.

ALLERGY: A HISTORY OF A MODERN MALADY

In the 1880s British doctor Morell Mackenzie noted that hay fever, with its tell-tale summer sneezing through the pollen season, went hand in hand with culture. Hay fever was an aristocratic disease, implying that the higher we rise on the socioeconomic scale, the more this allergic reaction tends to develop. Medical science at the time deduced that hay fever was the result of the stresses of modern life, the cry of an immune system struggling with its environment, and Mackenzie went on to conclude that the typical patient was 'a delicate, upper-class only child who subsequently developed into an emotionally and socially maladjusted adult'. In fact, although there are many influences involved, levels of urbanisation may be one of the strongest predictors for the prevalence of allergy in a population, both historically and in the future. This has been shown time and time again. Allergies are more likely to affect people who are more distanced from traditional upbringings. The history of allergy is perhaps a metaphor for our physical, emotional and cultural responses to our changing environment.

Cancer – A Nasty Trick on Our Immune Defences

Cancer is a wide range of diseases with a common characteristic: something goes wrong in the way the cells regulate growth, resulting in uncontrolled cell expansion. This happens deep in the cell at the molecular level, in the cell's genetic material.

Cancer is somewhat a matter of odds – an unfortunate by-product of the way evolution works. Large and complicated animals like us humans are vulnerable to cancer precisely because we are large and complicated. Each time our cells replicate, there is a chance of a mutation. Some of these mutations are benign, some are useful, but others are dangerous. Though it may seem flawed, over generations we pass on tiny mutations in our genes via this system. These mutations select out certain physical attributes, or phenotypes, to our offspring helping them better fit their environment. This is natural selection. But natural selection is not natural perfection. The more replications, the higher the chance of cancer. This is why our risk for some cancers increases with age. As previously mentioned, our immunity is our main cancer-surveillance system. But cancer is 'self'. It is part of us. So it can easily slip under the radar of our immunity cancer patrol. This is the cruel trick cancer plays on our immune defences. Some cancer cells present 'Don't-eat-me' signals on their surfaces, instructing our immune cancer killers not to attack our own tissues.

While there is much evidence that exercise and healthy living can reduce the risk of certain cancers, reducing is not preventing. You cannot remove all risk. Lifestyle can have a huge influence on the risk of cancer overall. My dad – an active man, a farmer, non-smoker, rare drinker, never overweight and having never eaten a takeaway in his life – was diagnosed 12 years ago with colon cancer, one of the most preventable lifestyle-associated cancers. With a colostomy and chemo (and a lot of faith in Mother Nature) my father survived the illness and has enjoyed good health since then. But, as I write, my family and I are anxiously awaiting more information on my dad's second cancer diagnosis.

Other than hereditary risk (mutations passed down in your family), and certain exposures and behaviours (like smoking), cancer is still pretty unpredictable. The older you get, the more likely it is that you will have it, and you might not even know it. And not all cancers have known lifestyle components. Even with the 'perfect' diet and lifestyle you are still at risk. But this doesn't mean that making healthy choices is futile. In fact, around 1 in 4 cases of cancer could be prevented in the UK each year largely through lifestyle changes including stopping smoking, maintaining a healthy weight, eating a healthy diet, enjoying the sun safely, protecting against certain infections and cutting back on alcohol.[10]

When it comes to cancer, the good news is that taking care of your health *will* help you get through treatment and recover in better shape than your less health-conscious peers. From the earliest days of my dad's cancer diagnosis, my mother instinctively stepped up her game to prioritise his wellbeing. Not a lifestyle overhaul, but paying attention to the little details: tweaks to his diet to ensure as much variety as possible and plenty of protein, prioritising sleep and reflexology to aid with the stress of such a diagnosis.

Metabolic and Lifestyle Disease

Metabolic syndrome – a cluster of metabolic aberrations that results in conditions combining diabetes, high blood pressure and obesity – is rampant. This clustering of risk factors doubles the risk of heart disease and increases that of type 2 diabetes five-fold. Research also links metabolic syndrome to an increased risk of several cancers. The diseases associated with metabolic syndrome can be chronic, debilitating and lethal. Even though each metabolic condition is different, these diseases share common features below the surface. Metabolic disorders are caused by a multitude of factors, but the immune system plays a role. Metabolic syndrome has strong inflammatory underpinnings often associated with deregulated immunity and reduced infection protection.

Something in our lifestyles triggers immunity danger signals that feed into metabolic deregulation. Many people with metabolic syndrome are overweight. Being over your healthy weight does put you at risk of metabolic derailment. In our modern environments it's easier to over consume calories, sending immunity slowly awry, as we will explore in Chapter 7. Studies suggest that this link reflects the influence of chronic inflammation and elevated levels of hormones involved in metabolic processes, like insulin. But weight doesn't tell the whole story, as we will see. The modern Western lifestyle hasn't just changed our diet, but every facet of our way of life. Epidemiological studies confirm that most people affected by metabolic syndrome have multiple common lifestyle characteristics or behaviours. These include smoking, poor diet, obesity and physical inactivity – all identified as leading contributors to overall mortality, working together to increase the risk of chronic conditions and other negative health outcomes in adolescence, adulthood and old age.

Brain Immunity

Until recently, the brain was thought to be cordoned off from the peripheral immune system. It's carefully tucked away behind a barrier to give it extra armour and prevent germs from getting in. But that extra barrier means that while most germy threats can't invade the brain, the body's immune system can't easily get in either. So it is reliant on its resident immune police-force cells called microglia. These cells patrol the brain. They swallow up invading germs, chow down on dead neurons and also trim the ends of healthy ones. (Neurons are specialised nerve cells in the brain, whose ends connect to each other, passing messages back and forth. These conversations help us to learn movements, store memories and much more, but the cellular connections change over time. Some become stronger. Others weaken. Eventually, some old connections aren't needed any more. So microglia step in as the neurons' personal barbers, pruning off old, less useful connections at the ends of cells so that new ones can emerge.) This single type of immune microglia in the brain exists in multiple flavours, working as versatile all-rounders, not specialists. So when they go rogue (following serious infections, injury or psychological trauma) they play a key role in mental-health problems and neurodegenerative diseases that rob people of their minds, memories and the ability to go about their normal lives.

Studying the relationship between mental health and the immune system is a hot area in psychiatry and neuroscience; so much so that new terms have been coined to define the field: psychoneuroimmunology describes the bidirectional communication between the brain and the immune systems and immunopsychiatry is the biological dominance of the immune system in the brain–body relationship. Inflammation was once considered a passive bystander in the health of brain function. Advances in science are beginning to unearth an unexpected role of the immune system in disease onset and pathogenesis. As we will explore, research suggests that for tens of millions of people, certain neurological disorders might be caused by a malfunction of the immune system. In other words, it is not 'all in the mind'.

A New and Dangerous Form of Inflammation

The discovery in the 1990s of a 'new' form of inflammation present in many, if not all chronic diseases goes some way to explaining the surge in our modern-day malaise. A subclass of inflammation, often termed chronic or 'meta-inflammation', is now used to define a form of low, smouldering inflammation. Unlike the short, sharp assault acute inflammation delivers, meta-inflammation is a long-term slow burn. It is not local to any specific area, but widespread, silently surging through your veins. New links between inappropriate inflammation and seemingly unrelated disease or disorders emerge in the medical arena on almost a weekly basis.

When inflammation shifts from short- to long-lived, there is a breakdown in immune tolerance and regulation, and major alterations in the normal functions of all our cells and tissues, increasing the risk for all non-communicable diseases. Chronic inflammation can also impair normal immune system functioning, leading to increased susceptibility to infection, tumours and poor responses to vaccines.

When someone is experiencing chronic inflammation, the causes are insidious, the 'danger' triggers are so subtle and the root issues can be difficult to pinpoint. Shaped through evolution, in an environment in which sedentary behaviour was minimal and calorie surplus was rare, immunity is not optimised to our present lifestyle. One of the first red flags for chronic inflammation was its link with obesity. But this slow burn can also be present without someone being overweight – it's just the cumulative effect of our overall modern lifestyle, comprising a plethora of psychosocial and environmental factors. Modern living has disturbed the balance of the immune system, setting off the 'danger' alarm bells in the form of an inflamed molecular cascade. This reaction is slowly undermining our health, constituting a cause or complicating factor in many diseases. We may not even realise, until slowly the cracks start to show.

Silent Signs and Symptoms

When you experience acute inflammation, the symptoms are pretty obvious: redness, heat, swelling and pain, all localised to the site of the injury or infection – like a sprained ankle or sore throat. But the symptoms of chronic inflammation are a bit more subtle and sometimes not even there at all. Signs that chronic inflammation is manifesting in your body include:

- Erratic moods, anxiety and depression (inflammation affects your entire body, including your brain)
- Pain (one of the key features is often painful, stiff or achy joints)
- Fatigue (inflammation places demands on your body, messes with your stress response and muddles up hormonal balance; it makes sense that your body is telling you to hit the sofa rather than the gym)
- No symptoms at all, but 'just not feeling right' (while symptomless chronic inflammation might seem pretty chill, it can still lead to all the risks and complications of chronic inflammation with symptoms)

There aren't, unfortunately, any standardised tests to measure chronic inflammation. A simple blood test can measure a liver chemical – C-reactive protein (CRP) – which, as part of the immune system, rises in response to inflammation. The result can help your doctor devise a strategy to lower your levels. However, the test doesn't show where the inflammation is or what is causing it.

When chronic inflammation symptoms do present themselves, they can manifest in a variety of ways, many of which can be mistakenly attributed to a lingering common cold or something else (hence chronic inflammation being so tough to nail down). Doctor visits are often related to the negative effects of stress or just feelings of being not quite right, unexplained or ongoing fatigue or niggling aches, pains or rashes. The red flags of chronic inflammation are often only acknowledged once severe symptoms that we can no longer ignore have taken hold.

While those symptoms all sound pretty awful, there's worse news: chronic inflammation has been linked to some very serious health issues down the line. Seemingly unrelated conditions like heart disease, type 2 diabetes, autoimmunity, allergy, some forms of cancer, and even neurodegenerative diseases like Alzheimer's share this deep immunological link.[11] This is because if it is continually 'rescuing' (i.e. fighting over a long period of time without resolution), your body's inflammatory response, left unchecked and unresolved, is universally damaging to your healthy cells, tissues and organs, which, in turn, leads to damaged DNA, tissue death and internal scarring. Chronic inflammation may be a recently observed phenomenon, but many of the related diseases are themselves not new. They are the same cluster of wear-and-tear diseases of ageing, but rates are increasing dramatically. And now they are happening more aggressively and at younger and younger ages in individuals living in industrialised countries who follow a Western lifestyle. But they are relatively rare in populations who adhere to more traditional diets and lifestyles, with environments and social stressors starkly different from those typically present in today's modern Western world.[12]

The idea that chronic inflammation – constant, low-level immune-system activation – could be at the root of many non-communicable diseases is a startling claim, requiring extraordinary proof. But is it cause or consequence? The details are still being fully worked out. The pathways of inflammation are complicated, difficult to define and manipulate, making unpicking the details challenging. But in 2017 a large-scale human clinical trial of more than 10,000 patients in 39 countries examined the effects of inflammation in heart disease. Researchers found that the immune system's inflammatory response is uniformly damaging and causal in the process, killing people by degrees.[13] For some conditions it appears that inflammation is a causal factor in this modern malaise. Until we accept that – and take action to change it – we will not be able to fight these lifestyle illnesses effectively.

The Drugs Don't Work

True, anti-inflammatory drugs exist. There are many types, including non-steroidal anti-inflammatory drugs (NSAIDs), many of which are available as common over-the-counter non-prescription drugs – ibuprofen, for example. These are powerful medications for a whole variety of acute pain problems, and when taken for a few days or a week they are very safe for most people. But they are not without their side effects. The Food and Drug Administration added new warnings about NSAIDs in July 2015. NSAIDs must be used cautiously in people with heart conditions and cardiovascular problems since recent studies have demonstrated taking prescription NSAIDs for just one week can increase the chance of heart attack or stroke in these patients,[14] not to mention the suite of more common side effects, including damaging the delicate lining of the digestive tract. And, alarmingly, they actually interfere with the resolution of inflammation, hindering the natural course of the healing process that our bodies are designed to do so well. In fact, these treatments may bring symptomatic relief, but a little-known fact is that they have minimal effect on chronic inflammation. Inhibiting these crucial resolution pathways over the long term induces physiologic dependence, which has been shown to produce a rebound effect, paradoxically promoting chronic inflammation further through a compensatory mechanism.[15,16] Constant use actually worsens some of the problems that people hope to ameliorate by taking these drugs. Arthritis is a prime example. NSAIDs provide short-term symptom relief, but long-term use accelerates disease.[17] This is why, and studies confirm this, that when it comes to chronic inflammation, the side effects might outweigh any benefits.

So when is it OK to take NSAIDs, and when is it not OK? This is a complex and controversial topic. If you have been prescribed these drugs, then you must follow your doctor's orders. Prescriptions aside, these drugs are ubiquitous and available without prescription, so now there's no guidance or accountability, and people are taking them for every imaginable pain or ache, even for emotional stress. They must be used with an eye towards the delicate balance that has allowed our species to survive this long. We need to start looking to better long-term ways to manage chronic inflammation, some of which we will explore in the upcoming chapters. I'm optimistic that holistic approaches to managing chronic conditions won't negate NSAIDs but may reduce how many we need to take.

Avoiding Snake Oil and How to Get the Best from Your Immunity

By now you should be getting the picture: immunity is more than those symptoms you feel when the seasonal winter lurgy rolls around. Immunity is the river that runs through every aspect of health and wellbeing. In my experience as an immunologist, it is so central to our wellbeing – survival, even – that its inner workings offer key lessons for living better. It keeps us alive. But it can also make us sick. We can no longer ignore the fact that our fast-paced, over-consuming, unrelenting lives are slowly eroding the balance of this carefully crafted system.

In my line of work I've heard it all. From chicken soup for the sniffles to copper bracelets for arthritis. Old-wives' tales, traditional beliefs and the positively bizarre. Each generation has their own 'snake oil' concoctions, potions or rituals promising miraculous cures and good health. Many are flawed or completely miss the mark, some are foolhardy and others downright harmful. That's not to say all alternative remedies are ineffective. Often, they seem to work for the people who promote them but may not be appropriate or are perhaps even dangerous for others. And, certainly, most won't be enough to keep you healthy all the time. For some, the biology of belief is enough. By this I mean the placebo effect, which is a wondrous thing (see p. 44). It shows us the immense power of our minds and the beliefs we hold, even though the impact that mindset can have on health is one of the least understood concepts in medicine.

On the other side of this coin is the rise of 'wellness'. This surging trend has ushered in a wave of enthusiasm for preventative self-care, but also propelled a cascade of misleading messaging, hyperbolic claims and a shift away from science. As we saw with the anti-vaccination movement that evolved from a discredited article in *The Lancet*, misinformation, compounded by confirmation bias, spreads faster and deeper than the truth. This means misguided choices and misspent money, both of which have the potential to compromise our health.

Everyone wants to hear the tactical. They would have me tell you what to do each day, which superfoods to eat and which supplements to take. So what's causing me to go against the grain and urge you to cast a critical eye upon all the messages of 'wellness'? To disregard the powerful marketing aimed at having you 'boost' your immunity? Some argue that 'the science just hasn't caught up yet'. But it never will. Because that's just not how science works.

PLACEBO EFFECT

A placebo is anything that seems to be a 'real' medical treatment – but isn't. It could be a pill, an injection or some other type of 'fake' treatment. What all placebos have in common is that they do not contain an active substance meant to affect health. They are used in clinical trials to determine the effectiveness of a new drug. Sometimes a person can have a response to a placebo, either positive or negative. For instance, the person's symptoms may improve or they may have what appear to be side effects. These responses are known as the 'placebo effect'. Your mind can be a powerful healing tool when given the chance. The idea that your brain can convince your body a fake treatment is the real thing and thus stimulate healing has been around for millennia. Now science has found that in the right circumstances, a placebo can be just as effective as traditional treatments.

I don't want to focus on lamenting times gone by or chastising the latest superfood. Life has moved on, and our health is better in some ways than it was in past generations. But I can't help but wonder if we have thrown the baby out with the bathwater? This book is not about quick fixes and cure-all remedies. For many, a mindset shift from sick-care to self-care means we've become more intentional about our diets, nutrition, exercise and lifestyle as they relate to health. As the evidence builds, conventional medicine can no longer afford to ignore its place either. It may be the best we have against the uncontrollable surge of NCDs that now plague our modern lives. Several scientific studies explored in the following chapters demonstrate that sustaining certain positive lifestyle choices lowers risk, and that those already diagnosed can have their treatment effectively reinforced by shifts in habits and the provision of strong support networks.

SOLVING THE PUZZLE OF BETTER HEALTH

Health, both physical and mental, starts and ends with your immune system – the guardian control centre behind all other body systems. Ultimately, the immune system seeks peace with the environment. Your goal is to create one that doesn't require your immune system to lose its natural balance.

Like the brain, the immune system is a learning system. It evolved to anticipate inputs from our environment. We can no longer ignore the fact that our fast-paced, over-consuming, unrelenting lives distort these inputs and are slowly degrading the balance of this carefully crafted system. Despite the fact that our world has changed – or, more correctly, because of that – it's more important than ever that we anchor ourselves to the things that came so intuitively to our ancestors yet present the greatest challenge in our modern-day 24/7 culture. For example, a sensible lifestyle, getting enough sleep, exercising regularly, eating healthy foods and cultivating the mental bandwidth to adapt and self-manage the constant pressures of modern life. The future is not conventional medicine *or* alternative medicine – it's a fusion of modern medicine with the intuitive tools of our great-grandparents.

For those of you who are curious about the forces within that can determine your wellbeing outside of your doctor's realm, but are nervous to venture into the sphere of Dr Google and online influencer quackery and tired of being told to juice celery, this book makes no claims for quick-fix cures but will provide you with knowledge and direction. Understanding immunity can help you navigate the long game for health. Instead of boosting your immune system: support it. Honour its finely honed balance – one that is so essential to feeling healthy and living well. In the following chapters I will share my own personal tips and tricks, teaching you how to incorporate modern versions of ancient traditions into daily life. Think of these as lifestyle tweaks and changes backed by science. This is in no way prescriptive, nor a magic bullet to health. Quite simply, it is a gentle nudge in a better direction.

CHAPTER 2

Living for Longevity

'Predictions are hard to make,
especially ones about the future.'
SIR PETER MEDAWAR, CBE (winner of the Nobel Prize in Physiology or Medicine (1960) for the discovery of 'immunological tolerance')

Playing the long game to health. I keep saying this to myself, while I grapple with the modern-day wellness culture, instant-health-promoting messages and unattainable body ideals. The older I get, the more I find myself using the long-game approach to health *and* believing it. The long game is having a long-term plan, long-term goals. It's about doing little things now, and every day, that set you up – well, for the long term. When you play the long game, you make conscious, considered, sustainable changes for the benefit of future you. This can apply to every aspect of life. But it is especially important for your immunity, as the governor of your health from birth to death.

But in today's hyper fast-paced world of immediate replies and instant gratification, the long game gets forgotten. We are distracted by daily demands. With too many choices and too much information, everything seems so important, and it becomes difficult to think about how our lifestyle today could impact our health in the future. We are caught in a landscape crowded with the business of health: juice bars, meditation retreats, detox diets, mindfulness apps and downward-dog-friendly Lycra – all while holding down a heavy life load. The result? We seek out the next quick wellness fix.

OUR IMMUNOBIOGRAPHY

Evolution is the slow change in characteristics shaping us slowly over millennia. By contrast, to survive day-to-day, we need to learn and adapt quickly to the immediate environment. In this chapter, I reveal how your immunity is in constant flux, adapting from cradle to grave. Starting with an immunity blank canvas when we are born (and perhaps even before), our personal immunobiography is moulded during childhood. It responds to the various environmental changes we are exposed to. From age and infection, to sex, attraction, gender and body size, our immunobiography is shaped by all our experiences. It is nurtured by our encounters and adventures, educated by our changing emotions and surroundings. It even determines our life expectancy, with age-related breakdown of immunity having serious ramifications for health and longevity.

Each life phase brings different health risks and challenges, underscoring the importance of the immune system's amazing capacity for lifelong adaptability. Our risk of poor health rises the longer we have been on the planet, particularly if what we plug into our bodies via lifestyle and environment is mismatched with what our immunity needs. In this chapter we explore how to self-manage our health given our ever-changing immunity and modern environment. Like me, you need to be in this for the long game. No quick fixes. Just tactics to help you navigate modern-day challenges.

SEX AND SMELLY T-SHIRTS

The way we smell is one of the most revealing markers of who we really are. It is determined by the same set of MHC compatibility genes (see p. 14) as our immune system – our personal chemical signature.[1] These pheromones are scents of attraction. And we are, in fact, surprisingly sensitive to them. All new-born mammals and their mothers can identify each other by smell alone within hours of birth. This is by no means perfect, and we don't make decisions based only on our sense of smell. However, having identified the right person, smell plays a very important role in that attraction and even sexual arousal. One study determined that men could tell if a woman was ovulating by sniffing a T-shirt she had slept in. The result? Men found the smell more pleasing when the T-shirts were worn around the time of ovulation.[2]

But why is this relevant to immunity? Just as our minds develop a feeling of identity, the immune system is completely individual. This is why it can reject organ transplants. We know that young mice tend to prefer the odour of their nest mates, but when they hit puberty, it's *vive la différence*. Studies in humans too show that we prefer the scent of a partner whose MHC compatibility genes are most unlike our own (sexy!). But when women are pregnant or on the oral contraceptive pill, their preferences revert, and they prefer the familiar odour of MHC-similar males. What this research is telling us is that we prefer mates whose immune systems are most different from our own. What might be the basis for these olfactory tendencies?

Preferring a partner with a dissimilar set of immunity genes might be an innate strategy to ensure genetic diversity. It may help us produce hardier offspring with the most effective immunity. The more genetically dissimilar a child's parents are, the more diverse their complement of MHC immunity genes will be. This gives them greater breadth of defence against any infection they are exposed to. It could also be an in-built strategy to reduce in-breeding and improve fertility (there is some evidence that MHC-similar couples are less fertile). The true chemistry of our relationships actually lies in our immunity genes, and while we haven't yet quite figured out how to use the power of pheromones, it hasn't been overlooked in the dating industry – there are now even matchmaking services all about the chemistry of our scent!

THE GENDER GAP

Although disease and ill health affect people of any gender, race and age, gender biases are particularly prevalent in our immunity. Not only do men and women respond differently to infections, but gender may have a knock-on effect on:

- the likelihood of getting a chronic disease
- the likelihood of developing allergies
- response to medical treatments
- the effectiveness of vaccines[3]

Unpicking this has largely been ignored by the medical and scientific community until recently. Female subjects have historically been excluded from clinical research and continue to be under-represented today. Medicine's gender bias has created an information deficit, unknown efficacy and deep ripple effects on human health.[4]

Why does this matter? Other than the obvious equality issues, women's historical exclusion from scientific research has serious implications for their health. Many more recent studies find unexplained fluctuations in the immune response in women. This could lead to an error in interpretation of clinical trials, known as a 'confounder', and might even affect how well they respond to a treatment like a vaccine. It's important for healthcare professionals to recognise that both sexual activity as well as the monthly menstrual cycle of women can cause natural fluctuations in blood-test results. If you're female, this could be useful to know if you are being treated for immune disorders or experiencing flare of inflammatory conditions. It may also help you with fertility issues.

MAN FLU IS REAL ...

Everyone seems to agree: men are drama queens about illness. But man flu is real – and so ubiquitous that it has been included in the *Oxford English Dictionary*: 'A cold or similar minor ailment as experienced by a man who is regarded as exaggerating the severity of the symptoms.' (True story.)[5] But how can men, who have evolved the innate biological ability to reach the top shelf and open jars, be weaker than women?

There is no arguing around science. Illnesses may actually affect men more than they do women. But this is not about over-dramatising symptoms. Evidence suggests they may in reality be suffering more. Symptoms may, objectively, physically affect them more severely. Globally, significantly more men die from infections.[6] The reasons why are complex and only partially known. Poorly understood reasons are probably at play: the survival of the species may mean men are harder hit by infection. Or could it be a sneaky evolutionary trick used by germs to enable their survival. Either way, men seem to have been singled out as the weaker sex when it comes to infections. Why? Well, women have developed multiple mechanisms to transmit infections. As well as the normal routes, women can pass bugs from mother to child during gestation or birth, or through breastfeeding. In this way, women are better vessels for germs to spread. As a result, they need a hardier immune defence against germs. Women might have evolved a particularly fast and strong immune response to protect developing foetuses and newborn babies. But this isn't the only explanation for man flu.

Anecdotal psychology claims that it's hard for men to express weakness. This means that when they are given a valid reason – they're ill – men really lean into it. There is also the role of sex hormones, the differences in how oestrogen and testosterone regulate inflammation and the production of antibodies. Oestrogen can activate the cells involved in antiviral responses, and testosterone suppresses inflammation. Hints that men and women deal with infection differently have been around for some time. In 1992 the World Health Organization hastily withdrew a new measles vaccine after it was linked to a substantial increase in deaths of infant girls in clinical trials. Yet very few studies assess men and women separately, so any sex-specific effects are masked. And many clinical trials include only men, because menstrual cycles and pregnancies can complicate the results. It's sort of an inconvenient truth.

... BUT WOMEN GET MORE AUTOIMMUNE DISEASES

If men are the weaker sex when it comes to fighting microscopic germs, then women must be blessed with immunological fortitude, right? Wrong. Nowhere does gender matter more than in the horrible lifelong diagnosis of autoimmune disease.

Autoimmune diseases have a strong female bias, with women three times more likely to develop one than men. Is a woman's reactive immune system perhaps leaving her more open to tipping over into autoimmunity?

Autoimmune diseases are complex and caused by many factors. As you read in Chapter 1, there needs to be a perfect storm of contributing genetics and a plethora of environmental influences. But this alone cannot explain why women are more likely to be affected. Layered on top are numerous ill-defined possible biological, social and psychological factors, and with autoimmunity on the rise, it is increasingly important to approach treatment and diagnosis through a gender lens.

Scientists know that genes associated with gender differences do have some effects on immunity. For example, they have been linked to different antibody responses to vaccines. What remains unclear is whether these variations are due to different genes being selected for in different genders or whether environmental triggers switch them on and off.

We do know that the female X chromosome contains a lot of genes involved in autoimmune disease. In contrast, the male Y chromosome has much less genetic influence on it. Generally speaking, women's immune systems are more robust than men's and mount stronger inflammatory responses during their reproductive years. So more than our genes and environment, the ebb and flow of key sex hormones across our lifespan shape us physically, emotionally and immunologically.

SEX AND IMMUNITY

Our sex hormones shape us from before birth to death in ways that we are only just beginning to unravel. These differences tend to appear after puberty and are particularly important immunologically in women during childbearing years.

Women's immune systems are in constant flux during the menstrual cycle. These female monthly cyclical changes can determine how susceptible women are to infection as well as to chronic diseases such as autoimmunity.[7,8] When we consider the nature of the key hormones controlling the menstrual cycle, this is no surprise. Oestrogen and progesterone can directly change the behaviour and function of multiple immune cells. During the first part of the menstrual cycle – the follicular phase, from the first day of the period until ovulation – a woman's immunity is more responsive and aggressive as oestrogen rises in the absence of progesterone. In theory, this means they are less susceptible to infection. But higher oestrogen does not necessarily mean better health. This higher inflammatory response can lead to a worsening of chronic inflammatory disease.

Conversely, following ovulation during the second half of the cycle – the luteal phase – the reactiveness of the immune system is temporarily curbed by rising levels of progesterone,[9] which is more typically immunosuppressive, blocking inflammation and protecting from autoimmunity. While women may then be more susceptible to infection, some may see not only a reduction in symptoms and flares of inflammatory diseases but also a delay in healing and recovery from exercise.

Testosterone, thought of mostly as a male sex hormone, is also made by women's ovaries, and offers some level of autoimmune protection, gently tapping the brakes on inflammatory immune responses. In fact, higher levels of testosterone are a key piece in the puzzle as to why men are better protected from autoimmunity.[10]

What's really surprising is that these monthly female fluctuations are exaggerated in sexually active women – yet another example of immunity in flux, responding to what we are doing. But why? It is a biological balancing act. When women are sexually active, their immunity is not only trying to defend them from anything foreign, such as an infection, but also has to tolerate the non-self foreign sperm to allow conception. Why? Well, regardless of whether women want to have kids or not, their bodies all experience the same biological drive. Put simply, evolution only really cares about passing on genes. So reducing immune defences when women are at their most fertile is the body's cunning ploy to increase the chances of pregnancy.

Why is it that a reduced immune response increases the likelihood of conception? The trade-offs between reproduction and immune response are a tricky problem for our immunity. As we learned in Chapter 1, the immune system attacks anything that looks genetically different from us. A foetus, with half of the father's DNA, is chock full of genetically 'foreign' material, so the maternal immune system has to be restrained throughout pregnancy to stop it from rejecting the half-foreign baby in utero. To increase the likelihood that a woman will become pregnant in the first place, evolution has developed a way of ever so slightly suppressing immunity in sexually active women. Amazingly, this immune suppression is triggered by arousal and sexual pleasure, and relates to how frequently a woman has sex. (We have yet to discover if there is any connection to the number of partners, but that would be interesting to find out.)[11,12,13] On the other hand, the more that men have sex, the more they improve their immune defences. Seems us girls can't catch a break.

All of this is immunity changing and adapting to social

behaviour. The immune system isn't passive, waiting around for an infection – it's intuitive, responding to environment and behaviour. Weirdly, celibacy is a legitimate immune booster! Only if you are sexually active will immunity make changes appropriately, deciding whether or not it needs to engage in these trade-offs.

The relationship between sex hormones and immunity doesn't end there. With huge fluctuations over a relatively short period of time, almost half of women who develop an autoimmune disease do so in the first year after pregnancy. Hormones are the key trigger for this unfortunate legacy of carrying a child. The spikes and dips of the postpartum roller-coaster or perimenopausal oestrogen (before the ovaries have called it quits) can drive the autoimmune disease wild. For this reason, it's important to pay attention to your cycles and seek help if you are concerned that your hormones are out of whack. But, of course, our sex hormones do not act alone. They are influenced by multiple factors, such as the level of stress molecules in our bodies at any given time, how much exercise we do and what we are – or are not – eating. And it's not just *how much* of any given hormone, but also their combination and their respective ratio that impacts on the development and the course of any condition.

As a woman transitions into menopause, the decline in female sex hormones results in a significant reduction in immune function, leaving it comparable to or even less robust than that of a man.[14] This can be a time when autoimmune symptoms improve, or due to a general decline in immune function, things can get a lot worse. It's difficult to tease out symptoms that are normal parts of the laundry list of menopause symptoms from those related to your immunity going haywire.

GIMME SHELTER – THE UNIQUE COMPLEXITY OF PREGNANCY IMMUNITY

You've read about how the immune system responds to sexual activity, but it is also front and centre when it comes to controlling pregnancy. Let's explore this relationship.

Immune cells are present at the site of implantation of the embryo. Scientists used to think that these immune cells were battling the foreign embryonic cells, which were, in turn, trying to suppress this immune response. If the embryonic cells had the upper hand, pregnancy occurred. But the battle continued throughout pregnancy. If this process was not successful, it was thought to lead to miscarriages or pre-term labour. But now we know this is not the case, and these clusters of immune cells at the embryo are helping develop a successful pregnancy, causing essential inflammation and wounding to the womb lining during the first 12 weeks. Over the next 15 weeks, the mother's immune system is suppressed. Regulatory T cells (the Tregs you met on p. 12) take the pressure off the default mode of the immune system, which is to reject the 'foreign' baby, forcing it to scale back its response, encouraging tolerance to the developing foetus. An aggressive immune system returns near delivery, when inflammation helps with the labour response. Through these carefully timed switches and continual negotiations, immunity promotes dialogue between mother and baby, creating the delicate balance required for life to occur.

The downside here is that a woman's immune system is running at a slower speed during much of pregnancy, so they are more at risk of infection. Mostly, even with the nastiest of colds, flus or tummy upsets, the baby is completely sheltered and won't experience any of the mother's symptoms. But a pregnant woman's immune system can sometimes overreact to an infection like flu or other triggers – such as stress, poor diet, chronic illness and allergies. This exaggerated immune response may lead to soaring levels of inflammation and may impact the unborn child's development right into toddlerhood. Certain infections can be transmitted from mother to baby and are particularly troublesome – for example, cytomegalovirus, toxoplasmosis and parvovirus. And some are harmful to both mum and baby, including syphilis, listeriosis, hepatitis, HIV and Group B Streptococcus. Getting one of these during pregnancy can have serious consequences. Do discuss any concerns about infection during pregnancy with your doctor.

True Human Tolerance

Immunological deviations from this carefully crafted immunity-pregnancy timeline are the cause – and consequence – of devastating pregnancy-related problems. If Mum and Dad are genetically different enough in their compatibility genes (no in-breeding here), this seems to stimulate specific pregnancy Tregs to tolerate the baby, guiding a healthy pregnancy to completion. In fact, during a normal pregnancy, these immunity regulators increase 100-fold and hang around in the mother for up to five years after the birth. The neat thing is that these Tregs are very specific for foetal molecules and return in subsequent pregnancies. They even suppress the anti-foetus immune attack more promptly, specifically and successfully the second time around[15] (there may be an upside to being the second kid!). So while being immunologically dissimilar causes us to reject each other's organs, in pregnancy it is a key stimulus for tolerance. Nature has a reason to try to reject an embryo in couples with similar compatibility genes. Being too similar would eliminate the ability to diversify, which would not be in the best interests of our immunity.

IMMUNE DIVERSITY IS GOOD FOR FERTILITY

Luckily, with the diversity of possible immune compatibility genes between us, rarely in the general population do two unrelated individuals have compatibility genes that significantly match. But it does happen. People with similar compatibility genes comprise a substantial proportion of the population experiencing extended infertility and recurrent pregnancy loss. To identify whether this is a cause of fertility issues, doctors are able to gene profile partners. This is not available on the NHS, but many private clinics offer testing. Hopefully, one day this will be accessible to everyone.

The issues we have explored not only affect the success of the pregnancy and the health of the unborn baby. Becoming a parent can change you fundamentally. And now we know that those changes take effect at the cellular level and define the structure of your inner defence systems. Over time, the physical composition of co-parents' immune cells shift to resemble those of their partner. Through habits, stress, diet and even sleep, parents eventually end up with more in common immunologically than identical twins, which is pretty remarkable for two unrelated people.[16]

Pregnancy can also impact the wellbeing of the mother. Many women with a diagnosed autoimmune disease experience remission of their symptoms during pregnancy, since immunity is dialled down and Tregs are turned up. Remission occurs in around two-thirds of women with autoimmune conditions during pregnancy. Whether or not this happens depends on how different the foetus is from the mother in terms of compatibility. The more it resembles the father, the more likely the mother is to go into remission due to the strong Treg response.

Autoimmune conditions were once seen as a contraindication for pregnancy. This is no longer the case, but there are still some risks and considerations. One is potential organ damage. For example, some autoimmune diseases cause kidney disease and high blood pressure. Another consideration is whether the woman has an autoantibody that could harm the foetus or has to take medications not advised during pregnancy.

HOW TO BUILD TINY SUPERHUMANS

As we've seen, our immunity is largely built in childhood, educated by our environment and shaped by our lifestyles, creating our own unique fingerprint for future health. The first five years really are *the* most important time for our immunity. No pressure then, parents. Now, whether you have children, you plan on having children, or you know children, the wonderful fact is that you have the ability to enable any kids in your life to be tiny superhumans, and many of these things can be helpful for adults too.

Before I go any further, though, let me remind you of something. None of us are perfect parents. And I'm in no way holding myself up as an example of someone who always gets it right. When it comes to the health of my twins, immunity is the main area where I question myself. It's what keeps me awake at night. It's as if I know too much. When you know a subject well, you can expect too much of yourself – a common cause of stress. This book is a guide, not a bible. Absorb what science tells us and use it to inform your approach. But don't punish yourself for not getting it right.

Is Illness Good for Kids?

Kids are pretty much an immunological blank canvas. Just as we all go to school, immunity needs to be educated (more on this in Chapter 3). We profit from our mother's immunity through maternal antibodies passed through the placenta during the last three months of pregnancy. These last a few weeks to months. The type and quantity of antibodies passed to the baby depend on the mother's own level of immunity. Premature babies are at greater risk because their immune systems are less mature when they are born, and they haven't had as many antibodies passed to them from their mothers.

After birth, if the baby is breastfeeding more antibodies are passed on in breast milk. But babies' immune systems are still not as strong as those of adults. Babies start to produce their own antibodies every time they are exposed to a germ, but it takes time for this immunity to fully develop. If you, like me, are a working parent with young children in day care, you might feel that the never-ending colds and flus are a 'natural' part of childhood. Perhaps you have even heard that they are vital for toughening up a developing immune system. This thinking is partly responsible for the resurgence of infection parties, where parents deliberately expose pre-school children to infected playmates. Their theory is that it's better to get the disease than to have the vaccine. Let's unpick these ideas and separate scientific fact from fiction.

Indeed, there may be a link between childhood early-life experiences and the possibility of developing allergies and asthma. But less publicised is the decade-long string of follow-up studies that disproved a link between infectious illnesses and protection from these inflammatory disorders. If anything, early illness makes matters worse. The link to a strong immune system isn't the disease-causing germs. It's early and ample exposure to harmless 'good' bacteria –

especially the 'old friends' you will be properly introduced to in Chapter 3.

Just as disease-causing microbes clearly bring on inflammation, these harmless micro-organisms appear to exert a vital calming effect on the immune system so crucial in the early years of life. A normal number of illnesses for pre-schoolers is up to eight per year. The figure may be even higher for children in day care or those who have older siblings. If a child experiences 12–14 illnesses annually, this is a hint that something might not be right – for example, a possible nutritional deficiency or something else going on that you need to talk to your doctor about. I'll discuss some of the key nutritional strategies to build your tiny superhumans' immunity on pp. 294–8.

Painkillers – A Safe Cure-All for Our Children?

All parents probably know the value of the sweet liquid (not mentioning any brands) – the paracetamol-based medicine that can reduce a high fever and soothe aches and pains. Indeed, many will admit to being semi-reliant on it as a safe cure-all.

Paracetamol is among the most common medicines given to kids. Once heralded as the go-to for preventing febrile convulsions in babies by reducing fever, the most recent research actually suggests there is no evidence for this, and it should only be used to make children more comfortable.

But doctors are now questioning whether we are too reliant on it. Children are taking three times more of the medicine than they were 40 years ago. But as a working mum of two I can understand it. When a child is unwell, parents want to do something about it; it makes us feel less guilty and takes the edge off stressful work–life juggling. But research suggests it can also become a bit of a habit, with parents addicted to the ritual of precautionary dosing up at the first signs of restlessness. Most studies suggest an association between paracetamol use in children and the development of asthma later in childhood, and babies given paracetamol just once a month 'are five times as likely to develop asthma' according to the news recently. However, several confounding factors in study design might contribute to this positive correlation, and without a prospective controlled trial, confirming this finding is challenging. Work from my old mentor at Imperial College London showed that paracetamol use in pregnant mothers didn't increase asthma in kids.

Don't get me wrong, paracetamol is the most widely used OTC and prescription painkiller worldwide and is often the first taken for a wide variety of conditions. It is generally considered to be safer than others that may be used higher up the 'pain ladder', such as NSAIDs or opiates. But in adults at least, we know that long-term heavy use is not without risks. And it's our dependence as parents that may be problematic too.

CHILDHOOD VACCINES

Routine childhood vaccines protect children from a variety of serious or potentially fatal diseases, including diphtheria, measles, mumps, rubella, polio, tetanus, whooping cough and others. Where our grandparents may have seen the horrible effects of these illnesses, we find them hard to imagine. If the burden of these diseases is uncommon – or even unheard of to you – it's because the vaccines are doing their job. As well as the routine vaccines offered, there are optional ones, like flu and chickenpox.

Vaccines are designed to generate an immune response that will protect us from future exposure to the disease. Individual immune systems, however, are different enough that in some cases a person might respond poorly. Meanwhile, germs are always trying to evade our various protection strategies. The flu is a key example of this. Like a Rubik's cube, the flu virus is constantly iterating its patterns, evolving and reformulating by shuffling its genes into limitless permutations. To many, the seasonal flu vaccine may be more like an annual gamble as scientists predict which strains will arrive, with variable degrees of confidence, and design a vaccine to match. Each year, public-health agencies essentially have to make educated guesses on this. But of course, educated guesses aren't always perfect. So the flu vaccine is not guaranteed to be 100 per cent effective. Does this mean that it's not worth having? Many would argue not. A flu vaccine carries very little risk (and cost), with potentially huge upsides. It can significantly reduce your risk of getting sick, and of you infecting others who might be more vulnerable, which is a huge benefit considering that flu can be life-threatening for some people.

The key concept when it comes to understanding the importance of vaccination is something called 'herd immunity'. This is how vaccines keep these pathogens out of the general population, protecting those with less hearty immune systems from getting deathly sick. The more people who get vaccinated, the less likely it is that an outbreak will occur. That's because vaccinated people can't infect others. Community immunity is a beautiful thing.

The call for a mandatory chickenpox vaccine in the UK may surprise many parents who consider the disease to be a mild one. For the majority of children, chickenpox is just a rather unpleasant illness and a nuisance for parents who have to take time off work to look after them. Most people develop it in childhood, but in adulthood it is far more severe. It is especially dangerous, maybe even fatal, contracted in late pregnancy or by people with an underlying health condition or receiving cancer treatment. So why reject the chickenpox vaccine? In those of us who are not vaccinated, after we experience chickenpox infection the virus stays with us for the whole of our lives, hidden in our nerves and under constant control by our immune system. Rarely, this immunity may wane over time, allowing the virus to reactivate, causing the disease known as shingles. Chickenpox vaccine is not as comprehensive in protecting us as the real deal, and vaccination is predicted to increase the occurrence of shingles in adults over time. This is because adults would no longer have their immunity silently boosted by exposure to chickenpox-suffering children.

Yes, vaccines do have risks, and this can rightly make parents concerned – hesitant, even. But we need to put the risk into perspective. Vaccines are rigorously tested and most routine ones now have long-term safety data. If current trends continue, the number of measles cases in 2019 will be the highest in decades. Although this gets little air-time relative to the potential risks, vaccines have benefits beyond helping to prevent the diseases they immunise against. Children who are routinely vaccinated against meningitis have the added benefit of a reduced risk for acute lymphoblastic leukaemia, the most common childhood cancer.[17] Despite the belief that acquiring an infection like measles 'naturally' is better, this is not always the case. In fact, the measles vaccine actually helps protect from other dangerous infectious diseases, whereas getting measles infection 'naturally' can be detrimental to your immunity with lifelong effect, leaving you vulnerable to diseases that otherwise might not have been a problem.

Tough Questions, Straight Answers

Here are some common vaccination myths and what science has to say about them:

Q. My child's immune system is immature – should I delay vaccines or just get the most important ones, so I don't overwhelm their immune system?

A. There's no proof that spacing out vaccines is safer. What is known is that the recommended vaccine schedule is designed to provide the greatest possible protection.

Q. Do vaccines contain toxins?

A. Vaccines contain small pieces of the germ they seek to protect against. These are either alive but inactivated, or killed. Vaccines do require additional ingredients to stabilise the solution or increase effectiveness. The dose is the most relevant factor, though. The volume of the solution contained in injection is so small that it is harmless. Worry about mercury in vaccines arose because some vaccines used to contain the preservative thimerosal, which breaks down into ethylmercury. We now know that ethylmercury doesn't accumulate in the body (unlike methylmercury, which does). Even so, thimerosal has been removed from all infant vaccines since 2001 as a precaution. Vaccines do contain aluminium salts to enhance the body's immune response, stimulating greater antibody production and making the vaccine more effective. Although aluminium can cause greater redness or swelling at the injection site, the tiny amount in vaccines has no long-term effect. It has been safely used in some vaccines since the 1930s. Trace amounts of formaldehyde (to inactivate potential contamination) may also be in some vaccines, but hundreds of times less than we humans get from other sources, such as fruit!

Q. Do vaccines really always work?

A. Research shows that 85–95 per cent are effective for most infections. But as you have learned, we are all so different. And so are disease-causing germs. So the effectiveness of each vaccine differs. Even our unique microbiota that you will meet in the next chapter can impact on your response to vaccines.[18] Flu is particularly tricky. The vaccine's effectiveness depends on the strains the public-health agencies pick – and sometimes they get it wrong. The 2018 vaccine was only 23 per cent effective at preventing flu.

Q. Would there be so much anti-vaxx info if there wasn't something to worry about? No smoke without fire ...

A. This is an interesting one. There is currently a real crisis of confidence in vaccines. Vaccine hesitancy seems to primarily stem from multiple factors including a mistrust of the science underlying vaccines. Rather than playing into the hands of anti-vaxxers, perhaps we need a different approach. We have a lot of data on the safety and efficacy of vaccines, and they have saved countless lives, but we are still learning about their broader effects. Taking steps to properly evaluate vaccines in randomised controlled trials could help alleviate suspicion and vaccine hesitancy.

Q. Are vaccines just a way to make money?

A. Pharmaceutical companies certainly see a profit from vaccines, but they're hardly blockbuster drugs. It's also reasonable for pharmaceutical companies to make money from their products. The costs of around 15 years' research and development to bring them to market are extremely high.

Q. Are the side effects worse than the actual disease?

A. Vaccinating children to protect them against life-threatening diseases can cause mild, short-term side effects, such as redness and swelling at the injection site, fever and rash. But the most serious risks, such as severe allergic reactions, are far rarer (1 in a million) than the diseases vaccines protect against.

THE MODERN ATOPIC PLAGUE

Prior to 1960, most of medicine did not regard asthma as common, and certainly not the epidemic it is today. Asthma, allergies and eczema have increased in all Western countries, a trend that became clear in the late 1980s. Since then, the number of cases has continued to increase, mainly in children, to epidemic proportions with serious consequences, bringing yet another worry to parents.

While we can't say for sure why allergy rates are increasing, they tend to run in families, so we know genetics is one culprit. This genetic risk factor is known as atopy, meaning that children where one or both parents have an allergic disease are much more likely to develop it. Further genetic indications are the fact that people with coeliac disease (a gluten intolerance – see p. 261) are three times more likely to have eczema, and relatives of coeliac patients are twice as likely to have eczema. Allergies are also considered a by-product of modern lifestyles, with pollution, dietary changes, vitamin D deficiency and less exposure to microbes tinkering with how our immune systems choose to respond to or 'tolerate' harmless things in our environment.

A diagnosis of eczema during childhood is often followed by food allergy, allergic rhinitis (also known as hay fever) and asthma, typically in that order. These related conditions are part of the 'atopic march'.

Eczema, Allergy and Asthma

Eczema is a chronic condition of itchy, dry skin, with periods of flares and remission (although most sufferers still have dry and itchy skin even without flares). Eczema nearly always begins in the first five years of life (but can affect people of any age). It is considered a touchstone for 'at-risk' kids on the atopic march. But fortunately, it resolves before adolescence in two-thirds of kids. Who continues to have eczema and who doesn't is impossible to predict. Children with atopic eczema do not follow a seven-year cycle. This is a myth, as there is no evidence that a change every seven years makes eczema better or worse. Even if children appear to 'grow out of' eczema, it can return during the teenage years or adulthood. But for the 50 per cent of people who have the altered filaggrin gene (known as a filaggrin mutation), eczema is a lifelong condition with an altered skin barrier for life.

Eczema is often called 'the itch you can't scratch' (though perhaps it should be 'the itch you *shouldn't* scratch'). Scratching can make the lesions worse and expose the skin to infection. It can also trigger other allergies. Healthy skin provides a remarkably effective barrier to our environment. Any parent will know that babies, classically messy eaters, tend to rub their food on their skin. If food accidentally gets into the fragile skin barrier of eczema-prone kids, it can sensitise them towards food allergy – and that's for life (see Chapter 7). In a very small percentage of cases, eczema is actually *caused* by sensitisation to foods, the most common being cow's milk, eggs, peanuts, wheat, nuts and fish.[19]

Avoiding Early Allergen Exposures – An Immunity Faux Pas

'Oral tolerance' is the term used to describe how the immune system appropriately recognises and tolerates food as something foreign but safe. For many kids this process goes awry, leaving them with horrible food allergies. Food-allergy development is down to the balance between the timing and amount of the potentially allergic food and, importantly, *where* (skin versus gut) our immune system 'sees' food. For example, breastfeeding transfers maternal antibodies and immune cells to the baby's gut, teaching the immune system to tolerate foods. Early exposure to a diverse range of foods when a baby starts eating solids has protective effects against allergy development. So avoiding certain foods in early life doesn't pay off. But this is where things get complicated for confused parents.

For around the last 20 years, doctors have been advising parents to avoid giving kids – from weaning up to a year old – the foods most likely to trigger allergic reactions, particularly things like milk, eggs, fruit, corn, peanuts, fish and soya, in the belief that this would prevent food allergies later on. Medical evidence now shows that not only was this incorrect, but that it may have contributed to the rise in food allergies. On the flip side, the number of worried parents who decide to restrict foods in their children's diets in the belief that they have adverse food reactions is increasing. Unfortunately, this could be setting them up to be on this diet for the rest of their lives, as restricting food but introducing it later in life can have the unintended consequence of actually inducing allergies.[20]

Restricting a diet without a clinical diagnosis of allergy is now not recommended and, in fact, delaying the introduction of these foods may increase your baby's risk of developing allergies. In countries such as Israel where kids have regular exposure to these potentially allergic foods, particularly peanuts, right from when they start eating solid foods, and especially while they are still breastfeeding, food allergy rates are significantly lower.[21,22] This even applies to babies at high risk of developing food allergies due to eczema. One study showed that babies with eczema who were given peanut butter between 4 and 11 months, continuing through to their fifth birthdays, were 70 to 80 per cent less likely to develop a peanut allergy. The current recommendation, based on international data, suggests that the best thing we can do to minimise food allergies is to introduce a wide variety of foods, including those that have been known to cause allergies, at around four to six months of age. Ideally, breastfeeding should continue for another six months after introducing food.

Food allergy is also more likely if your immune system 'sees' food through broken skin. For example, when a child scratches an eczema patch and food enters their skin, the immune system, which normally 'sees' food via the gut, 'sees' it in the wrong place. This confusion increases the risk of food allergy. Plus, scratching the skin seems to cause a chain reaction that affects the gut barrier, making it easier for allergens to be absorbed from the diet. Many parents shun conventional topical skin treatments, like moisturising emollients and steroid creams, since they don't get to the root of the problem. However, these are the most effective way to keep the skin moist and prevent cracking. This plays a crucial role in reducing the risk of a child with eczema developing serious food allergies.

What mothers eat during pregnancy is also important. This can directly influence the immune response of the unborn child and affect health through effects on the microbiota. Getting plenty of gut-loving fibre appears to be particularly important in this case. In fact, in an observational study of mother–infant pairs, all with a family history of allergic disease, higher maternal dietary intakes of resistant starch were associated with reduced doctor-diagnosed infant wheeze and allergy symptoms. Restricting a mother's diet of specific allergens during pregnancy and while breastfeeding, when a child is otherwise well, is not routinely recommended as a means to prevent food allergies. Breast milk is the ideal way to nourish your infant. It is least likely to trigger an allergic reaction, it is easy to digest and it strengthens the infant's immune system. Especially recommended for the first four to six months, it may possibly reduce early eczema, wheezing and cow's milk allergy. In fact, we now know that babies and children with food allergies are missing certain species of gut bacteria. These bugs play a key role in educating our immunity and teaching our Tregs to tolerate not only potentially allergenic foods, but other benign substances in our environments that can cause us allergies – like pollens and dust. It's still early days, but with our modern Western lifestyles wiping out precious communities of gut bugs, it's more important than ever to take care of them (see Chapter 3).

Asthma – An Epidemic No One Understands

Up to 70 per cent of kids with eczema subsequently develop asthma. Asthma is the most common chronic disorder among children and the leading cause of childhood hospitalisations. From 1985 to 2001, the prevalence of asthma rose 100 per cent. About 300 million people worldwide have asthma, 255,000 die from it annually and deaths could increase by 20 per cent over the next 10 years, according to the World Health Organization. Some of the apparent rise in cases may be attributable to doctors having got better at making the diagnosis. But increased reporting seems unlikely to account for all the new ones.[23]

People frequently blame asthma on air pollution, but its role is not entirely clear. Pollution does makes asthma worse in people who already have the disease, but it's not known whether it causes it in the first place. Breastfeeding exclusively reduces eczema, allergy and asthma risk. But bottle feeding alone won't *cause* them. A mother's diet during pregnancy and breastfeeding may play a role too. Studies of what women eat pre- and during pregnancy have identified a high fast-food intake (three or more times per week) and low fruit/veg consumption (three or fewer times per week) leading to kids with a greater likelihood of allergy.[24] Plus, children who are overweight or obese are at increased risk for asthma.[25] But what is vital is proper maturation of the immune system right from *in utero* throughout childhood. Maternal health, diet, stress and, importantly, the microbiota (see Chapter 3) are decisive for maturing immunity.

Another hallmark of the Western urban lifestyle is a low level of physical activity. Watching television is by far the most frequent childhood pastime. The question I have found puzzling is – is there something about children running around for hours that might protect them from asthma? Studies have shown that lungs are more likely to become twitchy if people go for a half hour or more without taking deep breaths. Taking deep breaths is what children do when they scurry around the playground or chase their siblings.

Like other chronic diseases, rising rates of childhood eczema, allergies and asthma cannot be explained by genes alone. They have causal relations to factors such as air pollution, obesity, diet and exposure to infections, antibiotics and allergens, including contact at very young ages. And they can have quite different causes in different people. The most strongly supported preventive measure is the avoidance of exposure to second-hand tobacco smoke. But how would that explain the increase, when overall, parents today smoke less than in previous generations?

If you are a parent of a child with allergies or have allergies yourself, the atopic march may sound like a harbinger of 'bad things' to follow. No one wishes this on their children, but can we do anything to prevent it? Fortunately, there are steps that may delay or possibly prevent allergies or asthma from developing, irrespective of any genetic risk. The difficulty is in differentiating between environmental factors that trigger the onset of allergies, eczema and asthma and those that exacerbate them.

IMMUNITY IN TEENAGERS

Of course, children eventually become teenagers. The teenage years can be a difficult time, physically as well as emotionally. Puberty is a major life transition. The sudden increase in hormones, rapid physical changes and the development of sex organs can put a lot of stress on the body. This giant developmental period brings a host of physiological, psychosocial and cultural changes. The process uses up lots of vitamins, minerals and energy, and we can't always buffer these stresses. And the coping skills teens develop may impact on their inflammatory responses – and overall health – by the time they are adults and later on in life.[26] This leaves them particularly vulnerable to infections as the immune system struggles to keep up. And coming down with an infection can affect schoolwork, exams and extra-curricular activities.

Adolescence is also a time of exploration when many may find themselves using alcohol or marijuana. Scientists have recently found indications that this may do serious long-term damage to the immune system. For example, teenage exposure to THC, or tetrahydrocannabinol, the main active component of marijuana, had severe alterations of immune responses in adulthood, with a clear switch towards pro-inflammation. This damage may result in autoimmune diseases and chronic inflammatory diseases, such as multiple sclerosis, inflammatory bowel disease and rheumatoid arthritis in adulthood. Other studies have highlighted how a specific part of the brain's immune system substantially increases the motivation to drink, particularly in teenagers, and this is linked to the desire to abuse alcohol later in life. Bear in mind that these tests were done in mice, so we don't know how translatable this is to humans. But the bottom line is that the immune system undergoes profound development in kids and adolescents; immunity can 'remember' previous exposures, stresses and changes, especially early in life, which can have important long-term consequences

THE IMMUNITY TATTOO TABOO

Immune cells sitting under the skin are ready to trigger a reaction in response to damage, such as that caused by a tattoo needle entering the skin and injecting pigment. The ink pigment is a foreign body, and the immune system responds and attempts to clear it. As part of this clean-up process, macrophages eat the ink, as they would with bacteria, a virus or other non-self material. But the ink gets trapped inside them as it is not easily broken down and removed as germs would be. As a result, the tattoo design is trapped in these skin immune cells. What happens next? As macrophages die, they release the trapped pigment. However, this freed ink is then engulfed by new incoming macrophages. This 'release–recapture' cycle continues indefinitely, thus ensuring the tattoo's permanence.[27]

A series of popular science articles have recently been touting the health benefits of tattoos, specifically their ability to boost your immune system. But is a little ink really the cure for the common cold? The answer, as is often the case, is not as simple as many of these headlines might suggest.

What the research showed was that people who were on their second or third (or more) tattoo experienced surges in a special antibody called IgA immediately following their inking session. IgA protects all the delicate barriers of the body like the gut and lungs. However, it's not clear how long these surges in immune strength last beyond the few minutes post-inking. So as long as the jury is out on whether or not tattoos can toughen you up against disease long-term, don't rush out to get one solely in the hope of staving off a cold. Getting a tattoo may have some unexpected complications if your immune system isn't up to scratch. That includes people with underlying health conditions or those taking immune-suppressing drugs (often given after an organ transplant or to treat autoimmune conditions). Others who may have weak immune systems include those with chronic long-term conditions such as diabetes.

BODY WEIGHT IS AN IMMUNITY ISSUE

'Fat' is a loaded word. Modern-day society (wrongly) deems it unsightly and associates having too much of it with many health problems. Although this is partially true – being overweight is a risk factor for many inflammatory and metabolic diseases, even certain cancers – fat also has a good side: it's a critical component of our immunity. Contentious and contemporary, this means body size is also an immunity issue.

Fat (adipose) is divided into white and brown fat. Brown fat cells express high levels of thermogenic genes and help maintain body heat by burning calories. White fat is what we typically think of as 'fat' – it's the cells involved in whole-body energy and storage of excess energy. This is the fat associated with our immunity. Some in my field may even call white fat cells primitive immune cells. Take belly fat, for example. This is a type of white fat called visceral adipose tissue (VAT). VAT is laced with immune cells filtering the fluid that leaves our digestive tract, sensing microbes, damaged cells and unruly inflammation. VAT and immunity work together, with VAT providing energy for immune activities. Immune cells influence the biology of these fat cells themselves, altering their ability to store more lipids in favour of antimicrobial function. Fat is also home to rapid responders – the immune memory cells you met in Chapter 1, which are the first on the scene of anything untoward.

So it turns out that having a healthy body weight actually contributes to a stronger immunity. Get this delicate weight balance wrong – in either direction – and you could be opening yourself up to health risks. Unfortunately, despite VAT being key to infection fighting, having too much of it is not a surefire route to strong immunity. Maintaining a healthy body weight that is right *for you* is therefore the best way to go.

A series of studies examining the link between inflammation and insulin resistance in diet-induced obesity, showed that immune cells – called macrophages – reside within the white fat tissue to maintain body weight. Macrophages living in VAT are more likely to become chronically activated and produce inflammatory cytokines under certain circumstances. This is an adaptive effect in response to eating too much, which places a stress on our fat cells as they struggle to store excess energy. This 'danger' recruits more inflammatory immune cells, further aggravating inflammation. It also instructs the fat cells themselves to malfunction.[28,29] So fat cells become engorged with fat, increasing in size and number, which, in itself, causes a stress alarm and perpetuates the inflammatory scenario. So persistently enlarged fat cells mean that your body is always in a state of inflammation; your immune system is permanently 'switched on'.

Much less research has been done on too little body fat. That said, we do know that fat is an important storage site of memory cells.[30] Not just any memory cells – a group of super-immune cells with enhanced function. These cells remember past infections, providing you with enhanced protection. Too little body fat and you might not have enough space for memory cells. Too much body fat, however, can take up valuable immunological 'space', leaving little room for fresh new cells. So it's quite important for our defences to carry enough body fat, but not too much. What does 'enough' body fat actually mean, though? There are no blanket recommendations, but some objective measures to guide you include sufficiently developed muscle mass (see p. 203) and a body-fat percentage in a healthy range (10–20 per cent for men; 18–28 per cent for women). Then there is also thinking about your calories – consuming neither too many nor too few – being active rather than sedentary, and fuelling for your level of activity. It is generally accepted that the belly is the most dangerous place to carry excess body fat, so waist circumference has recently become a popular extra measure to determine whether you have a healthy level of body fat.

Why does body composition play such a key role in our immunity? Some of the biological mechanisms that are unhelpful in modern society are actually those that kept us alive in prehistory. People who carry more fat may have fared better during famine, for example. In evolutionary terms, a sudden or rapid weight loss could be a more immediate threat to survival – a signal of stressful danger – activating inflammation, while slowing down the more specific adaptive arm of immunity. Your immunity is needed to help turn white fat into the much healthier brown fat. We know that people who are obese often have sluggish immune systems and a lower amount of this white-to-brown-fat conversion.[31]

We know there are long-term risks for carrying too much weight. There is little disagreement that extreme obesity is a bad thing. But there is debate around how obesity is defined, and whether we should consider simply how much fat someone has or look more closely at where that fat is located in the body. But obesity can also be an independent risk factor for developing disordered patterns of eating via an impact on body esteem. So we need to work hard to break the stigmas that come with it. And it is not just your body fat – muscle mass is important for immunity too, as we will explore in Chapter 6.

GREY DAWN – LET'S TALK ABOUT AGEING

The huge privileges of modern living and, with it, convenience culture – free of the challenges our ancestors faced – are actually, very slowly, killing us, reducing our health span and enjoyment of life.

Recent studies on this ageing paradox shine a light on the immune system. Health span and lifespan are determined by our immunity. Immunologist Dr William Frankland, who, at 105 years of age still works in immunology research, publishing regularly in scientific journals, is an inspiration. He has got me thinking:

What enables some people to be so vital in old age?

Are they mere outliers, or can anyone, potentially, reach a ripe old age in good health?

The quest to live for ever, or to live for great lengths of time, has always been part of the human spirit. Longevity is a function of lifespan and health span. Lifespan is pretty easy to define. It's the number of years you live, your chronological age. Health span is intuitively obvious, but a bit harder to nail down. For simplicity, let's agree that it's a measure of how well you live – your biological age. While chronological age is quite easy to measure with a high degree of precision – at the time of writing, I'm 37 years, 11 months – biological age is less so. One without the other (long lifespan with poor health span, or short lifespan with rich health span) isn't what most people want.

Ageing is perhaps the greatest challenge that healthcare systems will have to deal with this century. But is ageing itself a disease? Greek doctor and philosopher Aelius Galen (*c.* AD 121–169) set the conceptual framework for our understanding of ageing. He defined disease as an abnormal function. Since ageing is universal, it cannot, he reasoned, be a disease. Later variations on this argument led to the same conclusion: if ageing takes place in everyone, and disease occurs in only a part of the population, then disease and ageing are not synonymous. The former should be cured and the latter endured, or perhaps celebrated.

In the past century, science and medicine have extended life expectancy; now 70 is the new 60 – a significant social transformation. But at the same time, the odds of living a long and also healthy life are, perhaps, not moving in the right direction. We are faced with an ever-accelerating pace of life, relentless stress, pollution, circadian disruption, malnutrition from overconsumption of a calorie-rich, nutrient-poor diet and sedentary jobs. Counterpointed with intense gym sessions and reduced exposure to micro-organisms, this leaves our immune systems in turmoil.

Ageing is the predominant risk factor for most chronic diseases and conditions that limit health span. The World Health Organization (WHO) released a report in 2015 stating there is little evidence that older people today experience better health than their parents did at the same age. We are living longer, but not better.[32]

The great triumphs of public health in the 19th and 20th centuries in sanitation and understanding the causes of infectious disease produced what scientists call an epidemiological transition. In other words, a sharp fall in deaths from acute infectious disease in early life and a rise in chronic inflammatory and degenerative conditions – in part because more of us are surviving into later life. Meanwhile, age-related diseases are affecting people at an earlier stage, with some young people ageing three times faster than they should.[33] The success story of longer lives is a worthless prize if quality of life is compromised because of poor health and a loss of autonomy.

Our aim is a great health span. But what will help us achieve this? To answer this question, let's start at the end of the story. Consider centenarians who are genetically gifted with a long and healthy life. Because life expectancies worldwide are below 100 years, the term 'centenarian' is associated with longevity. But centenarians often die of the same type of diseases as the rest of us. The difference is that these diseases seem to have a delayed onset.

The Ageing Immune System

Health span is moulded by our immunity, particularly cumulative inflammatory damage. This doesn't age us chronologically. From birth we are in a slow state of immunological decline, with under 40 considered young and over 60 considered old. But the ageing of our immune system is not just a straightforward decline. Rather, it remodels itself as it grows biologically older and more tired. This has been shown in identical twins, who become immunologically more individual as they age.

There is still a bit of a mystery about ageing immunity. What we do know is that as the immune system ages, the specific adaptive immune system deteriorates. Much of this is down to the thymus gland shrinking. The thymus is the backbone of the immune system, producing those master controller T cells, but over time the thymus turns down the production of healthy new T cells. We lose regulation via those all-important Tregs, which protect us from the inflammatory collateral our own immunity can wreak upon us. We may notice that vaccines don't work so well or we seem to get sick more often or struggle to shake off infections as we grow older. And risk for cancer rises exponentially because of our declining super-important immune surveillance (see p. 11). In fact, an ageing immune system may be a better predictor of cancer than looking for mutation in our DNA.[34] On the flip side, ageing increases our all-guns-blazing innate inflammation – the single-most important driver and predictor of ageing and age-associated disease! Ageing-associated inflammation even has its own special term: 'inflammageing'. This sucks up our immunity's precious time and resources. When we do get an infection, it's like fighting a battle in the desert.

Ultimately, the longer we spend on the planet, the higher our risk for many things, not least our immunity. The information theory of ageing assumes that we lose health and function as a result of our housekeeping and repair mechanisms getting worn out repairing the accumulation of genomic damage in our cells. Our genes carry our body's genetic information code. And telomeres are the little protective tips at the end, like on the end of a shoelace. They are made of repeating short sequences of DNA sheathed in special proteins. As we age, they get shorter. This has long been thought to be the strongest predictor of how healthy someone will be in their old age. Inflammageing erodes our chromosome-protecting telomeres and, at the same time, wreaks havoc with our delicate cells and tissues, demanding that our repair-and-protect mechanisms work harder than ever.[35]

So, once again, immunity is woven into the fabric of our health and the emerging predictor of longevity. Everything we know about ageing tells us it's not a programmed process. It's a question of our genes (about 25 per cent), with the rest being down to our exposome – the totality of our experiences, emotions and environment across our lifespan. Think of it in terms of getting a new car. How long it lasts will be partly due to the model and partly to how you treat it.

Immunity and Oxidative Stress: Striking the Balance

'Oxidative stress' is one of the latest buzzword terms, used to describe what happens when the biological balance between our antioxidant defences and free radicals tips unfavourably towards the latter. Our cells produce free radicals from the myriad normal metabolic processes. Our cell's battery packs – the mitochondria – run our metabolism, turning food into energy with free radicals produced as a by-product. But these are unstable and highly reactive, attacking us and causing damage to our tissues. Oxidative stress is a normal process and, in general, the body is able to maintain a balance between antioxidants and free radicals.

The effects of oxidative stress vary and are not always harmful. In fact, having too few free radicals around results in reductive stress, which is also detrimental to health. As you will read in Chapter 6, oxidative stress plays an important role in the benefits of exercise. But it does make a significant contribution to ageing and age-related diseases such as cancer. (Increased oxidative stress induces DNA damage and cancer-causing mutations, making our own repair-and-protect machinery work harder.) Several things can accelerate it – smoking, poor diet and lifestyle factors (covered in later chapters), as well as pollution. Inflammation also causes oxidative stress temporarily as the immune system fights off infection or perceived danger and reacts with free radicals as part of its weaponry. Oxidative stress can also trigger the inflammatory response, which, in turn, produces more free radicals that can lead to further oxidative stress, creating a cycle of damage and decline.

Want to Live for Ever? Get Rid of Your Zombie Cells

Just as a car eventually rusts, we become more 'oxidised' as we age. Uncontrolled oxidative stress can damage cells, proteins and DNA, and is linked to many diseases of ageing including autoimmunity, metabolic disease, some cancers and neurodegenerative conditions. The imbalance between free radicals and antioxidant defences during oxidative stress is known to make immune cells in particular become what's called 'senescent'. Senescent immune cells look very different from healthy cells.[36] These move around, but pretty much like revived old corpses – or zombies. They are prohibited from reproducing but pump out inflammatory SOS signals and a cocktail of other foul factors indiscriminately in an attempt to recruit other healthy cells to their rescue, often in vain and while poisoning surrounding tissues. These signals create an internal noise of inflammation that causes damage to tissues. This, in turn, further activates the immune system to repair the damage.

As we age, the number of immune cells that are fated to be zombies increases. And since they like to hang out in our white fat tissue, the more excess fat we carry the more zombies we can accommodate. Depleting old 'zombie' immune cells can increase lifespan by 25 per cent, at least in experimental animals. This has been shown to delay or alleviate everything from frailty to cardiovascular dysfunction to osteoporosis to, most recently, neurological disorders. Poor diet and lifestyle, stress and sedentary behaviour have all been shown to promote senescent cell accumulation. How can we rejuvenate the immune system and delay the onset of age-related diseases? Antioxidant-rich phytonutrients (see p. 246) play a key role in protecting from zombie cells.[37] Plus, fasting can be one of the best ways to clear out these broken immune cells. But it's not as simple as just popping a pill or taking a few days off from eating, as you will read in Chapter 7.

Older, Less Wise and Definitely Exhausted

After years of fighting, our immune cells become exhausted. More and more of them are now 'memory cells'. Immune memory gets wiser with time, accumulating memories of things we have been exposed to and vaccines we have been given. In theory, this should do more to keep us well. But this memory takes up a lot of immunological space. The bone marrow – our immune-cell factory – slows down with age as our bone-marrow stem cells sustain oxidative damage to their DNA. In a nutshell, the toolkit needed to set up new immune-cell armies in response to infections is pared down in elderly people due to overcrowding. So their immune systems may only be able to target old strains of viruses and bacteria with their outdated arsenal. This is known as immune exhaustion.

So when we encounter a new challenge, our ageing immunity is just not up to the challenge. One thing that takes up a lot of immune memory is long-term viruses. In fact, researchers first discovered this by studying patients with chronic viral infections like Epstein-Barr (EBV) and cytomegalovirus – part of the herpes virus family. These are among the most common viruses in the world, yet most of us have never heard of them. Even once infected (and most of us eventually are), many people experience few, if any, signs and symptoms. These viruses are problematic because they are persistent inside our bodies, hanging around for a lifetime. And keeping them in check puts so much demand on our immune cells that they actually accelerate ageing of the immune system by taking up valuable immunological space. This leaves less room for fresh new cells and those all-important Tregs. Chronic infections like EBV keep the body in a low-level state of SOS, and they sneak in when we are distracted by life – stress, life changes, etc. There is, of course, no good time to deal with a disease, but infections like EBV are opportunists, kicking us when we're down.

Slowing the Immunological Ageing Clock

When it comes to an ageing immune system, it's a two-way process: ageing changes our immunity; and unruly immunity drives ageing. The immune system becomes slower to respond, increasing your risk of getting sick. Vaccines may not work as well or protect you for as long as expected. Autoimmune disorders and cancers are more likely to develop. Your body may heal more slowly as there are fewer immune cells to bring about healing. As part of this process, the immune system's ability to detect and correct cell defects also declines.

It is important to understand the two-way ageing process. But don't buy into the idea that your health *has* to decline as you get older. We all take the journey with its many surprises, but we can buffer the bumps and make it more palatable as we go. We might not want to live to 100, but there are things we can do to keep our vitality. So how can we prepare our bodies for a longer, healthier life?

A major collective step to increase health span as well as lifespan is to support science and research into ageing. On an individual level, there are areas of our lives where we can take steps to keep our immune system in balance and decrease the risks from immune-system ageing. In the next chapter, we will look at the microbiome, stress, sleep, exercise and food – along with actionable insights to improve our health.

CHAPTER 3

Have Your 'Old Friends' Got Your Back?

'We are not individuals. We are ecosystems with microbial partners that are involved in the development (particularly in early life) and function of essentially every organ, including immunity.'

GRAHAM ROOK, Emeritus Professor of Medical Microbiology, University College London

As we have seen, our immunity is about more than the genes we are born with. Responding to, learning from and adapting to our environment, it is evolving from the moment we are born. And it does not evolve in isolation. Much of this environmental interaction is conducted by the rich and dynamic ecosystem of microbes that live on us, in us and around us – collectively known as our microbiota. Yes, we are superorganisms.

Based on current estimates, every one of us is home to a community of 38,000,000,000,000 (that's 38 trillion) microbes – our microbiota – accounting for half of each of us (by cell count). Much as we fear germs, the bugs comprising our microbiota are actually our biggest health allies. Our lifelong task of trying to balance our immunity means relying heavily on certain exposures to good germs for proper calibration. Why?

Well, when we look through evolution, our immunity developed through times of great peril, interacting with germs of all kinds – good and bad. When operating optimally, this immune–microbial alliance creates a dialogue that underpins our immunity, selecting, activating and terminating many of its various components as we need them, helping us distinguish self from non-self, danger from non-threatening. It determines how susceptible we are to infections and our likelihood of getting an autoimmune disease. This dialogue even controls how our brains function.

Our relationship with these microbes is what puts the 'super' into 'superorganisms'. But this relationship is changing in the face of wider social and environmental shifts. Our sanitised modern lives mean that our immune systems come into contact with certain microbes *less often* or *later* in life than before. The modern immune system's training and development is stymied by the gradual absence of its customary microbial targets. With nothing constructive to do, this leaves it crazily spinning its wheels. The result? Soaring rates of childhood allergies and autoimmunity, even heart disease and cancer. This effect may even account for many cases of autism. When we start to piece all this together, we begin to understand a human as one large creature made up of billions of smaller microbial ones. These, mostly gut-dwelling, microbes have been with us since the dawn of time and are as important to us as a limb or an organ. They don't act in isolation but in diverse communities. Think of them as factories of bioactive products that set our immune rheostat, educating and shaping the development of our immune-system development. Ultimately, this determines our disease-susceptibility patterns later in life.

Everything we look at is covered in microbes – we just can't see them. And 99 per cent of them won't hurt us. The immune system is like an athlete: to become strong and adept, it needs training and practice, and these 'good' germs are its friends and coaches. As you read earlier, our old friends are the vital microbial exposures, coaching and educating the immune system – not the colds, measles and other childhood infections, but rather the many harmless microbes and dirt, dust and general miasma around us from birth. Science has now shown, beyond a shadow of a doubt, that the microbiota plays a significant role in health.

OLD FRIENDS ALLIANCE

We all have microbial old friends with which we form a lifelong health alliance. Mammals, birds, fish and even plants have them. And then there are the viruses that teem inside those germ cells and ours. Where on our bodies are these old friends in residence?

They exist almost everywhere on and in the human body that we have looked, although some areas contain larger concentrations than others. They are clustered around the barriers of our bodies (gut, skin, lungs, urinary tract), reinforcing these boundaries, the good guys out-competing the bad guys: on the skin, bacteria promote healing; in the vagina they guard against unwanted yeast; and in the mouth they break down food and protect teeth and gums. But that is just the tip of the iceberg. There is possibly no organ system in our body that the microbiota isn't connected with (gut–brain axis, gut–muscle connection, even gut–gonad axis) and we are starting to find bacteria in places we never expected (placenta, brain, eyes). No single part of our bodies is sterile, with microbes making up to 2kg of our body weight. I want to tell you that these bugs living on and in us, indisputably, set the lifelong course of our immune function. In fact, living happily together with these helpful germs has been a major influence in crafting our immunity over millions of years.

The biggest microbial load, the gut, sits at the core of it all, connected to and influencing everything else, not least our immunity.[1] By now, it's a familiar fact (and yes, it's true) – indeed almost 70 per cent of the entire immune system resides in the gut. Throughout life, we are constantly throwing all sorts of potentially problematic things into our mouths. The essential task of the gut immune system is to maintain a balance between immune reaction and tolerance. It is essential that this tolerance – called oral tolerance – is established from a young age. To put it simply, our guts contain billions of bugs with types from hundreds of different species that collectively contain over 3 million genes, putting our own meagre 25,000 to shame.[2]

WHAT LIVES INSIDE YOU? A QUICK GUIDE TO YOUR OLD FRIENDS

The terminology around good microbes can be confusing. Here's a guide to help you navigate the ins and outs of our body bugs:

FLORA, MICROBIOTA, MICROBIOME

These all refer to the different micro-organisms living on and in us. The terms tend to be used interchangeably, but each actually means something slightly different:

- Microbiota, or flora, usually refers to all the specific micro-organisms that are found within a specific environment including bacteria, viruses and fungi. This means that there are localised differences in the microbiota of each person, depending on where in the body the microbiota resides, e.g. gut microbiota can be radically different to skin microbiota.

- A microbiome is the collection of genetic material from all the microbiota in a particular area, such as the gut. The Human Microbiome Project found that there may be more than 8 million unique genes associated with our various microbiota. This means that our total microbiome may give us more than 200 times more genes than reside in our own cells. These microbial genes are essential to our development, immunity and nutrition.

SYMBIOSIS
This refers to the close and long-term biological interaction between two organisms. Symbiosis is now considered an important selective driving force behind evolution.

EARLY IMMUNE EDUCATION

Our individual microbial fingerprint is the sum of our age, gender, medical history, geographical location, habits and diet. It's also shaped by physical activity, stress, sleep and a myriad other undiscovered factors. If you want to take advantage of the immune-nourishing effects of your microbes (and you definitely should want to), your first task is to accept that there are limits to what you can do because the roots of your health and delicate immunological balance lie in how you are born and nourished during the first five or so years of life. This period of critical immune education sets the immunological tone for the rest of your life.[3]

The birthing process sets off this most radical transformation as a tsunami of microbes colonises the (relatively) sterile baby as it enters the world.[4] But subtle calibrations may be happening upstream of this. Receiving these foundational microbes from your mother at birth, known as seeding, educates and shapes your immunity, impacting your health decades down the line. This early immune education happens predominantly right in the microbes' front yard: the gut. Microbes living in the gut coax immune cells to embrace these good germs, rather than spark an immune response.[5] As we've seen, immunity is not a binary on/off. The microbiota is similar in concept to a rheostat, with these germ friends playing a significant role in determining the inflammatory set point. Tuning the immune rheostat to a desired set point happens in the first five years of life through exposure to microbial 'old friends'. This 'licenses' the threshold at which immunity is triggered. For example, this vital symbiosis alters our resistance to infections and is important in toning down the type of immune responses more associated with allergies.[6] In fact, the task of developing a vital proportion of the regulating Treg cells are so important in preventing autoimmunity is outsourced to the microbiota. These gut guardians can even reverse food allergy, at least in experimental studies.[7]

This immune education doesn't just stay in the gut. Educated immune cells travel all over the body. The result? Any major alterations in the gut can send the whole system crazy, permitting inflammation to wreak havoc unpoliced. Without these early microbial guests in the first years of life, immunity as we know it would not exist. Interfering with this crucial process early in life is akin to fooling around with a vital organ like the liver or spleen.

Babies born vaginally primarily acquire the microbes that inhabit their mothers' vagina and bowel. Those delivered surgically acquire them mainly from the mothers' skin and the birth environment.[8] Caesarean deliveries can distort or inadequately populate the micro-organisms in a baby's gut, and these differences persist in children until at least seven years of age,[9] meaning they are significantly more likely to develop asthma, autoimmune conditions and even leukaemia.[10] The detail behind these associations is immensely complex and not yet fully understood. It's not quite as simple as saying C-section-born children are destined for worse immunity. As a mother, I know that vaginal delivery is sometimes not an option. But you can still promote the development of a hearty microbiome for your baby, as you will read below.

EVERY BABY NEEDS A SUGAR MAMA

After their route of entry into the world, what's the next thing to influence the microbiota of a new-born baby? Sure, microbes are picked up from adoring visitors or a lick from the family dog. But the biggest conduit for passing bacteria to a baby is breast milk.

Human milk is a particular microbial marvel. Once considered sterile, breast milk is, in fact, a creamy bacterial soup. It even contains microbes from the mother's gut that travel to the breast in preparation for setting up the new child's microbial ecosystem. These early inhabitants influence and select for bacteria that follow, leaving a footprint that can be detected even in adulthood.[11]

Breast milk also conveniently comes packed with a sweet food source, not for baby, but for their microbes. Human milk oligosaccharides are sugars found only in breast milk.[12] In fact, they are the third-most plentiful ingredient in human milk, after lactose and fats. Their sugar–fibre structure ought to make them a rich source of energy for growing babies, but babies cannot digest them. Why would a mother expend so much energy manufacturing these complicated chemicals if they were apparently useless to her child? These breast-milk sugars pass through the stomach and the small intestine unharmed, landing in the large intestine, where most of our bacteria live. They are not food for babies, but food for bugs, specifically designed to feed the growing gut biome and encourage correct development of the immune system. In other words, the microbe's full beneficial potential is unlocked only when it feeds on breast milk. As our bugs digest these breast-milk sugars they release a suite of powerful metabolites that both educate the immune system and calibrate the inflammatory set-point. As we mature through childhood, our food sources diversify, continually shaping our microbiota.

FORGOTTEN FIBRE

If 2019 was the year of gut health, then it's definitely been fibre's turn to hit the 2020 limelight. If our immunity is beholden to the health of our microbiome, it follows that we need to feed those gut bugs well. After birth and breastfeeding, nothing matters more than giving them the right food. Fibre is good for you and now science knows why: fibre not only helps keep our bowels regular but provides some great fodder for your microbiome and might just be the lifesaving food 90 per cent of us aren't eating enough of. Although not often considered an important 'immune booster', through feeding our microbiome, fibre plays a vital role in the overall capacity of the immune system to do its job. The preferred food for our gut bugs, dietary fibre is the most accessible tool to nurture our gut microbiome and put our immune system on the right track from the start.

What we eat can change the balance of microbes in our guts – and as their relative numbers change, they secrete different substances, activate different genes and provide us with different nutrients. Our modern-day diets are inherently lower in fibre than our ancestors'. This has led to a loss of our ancestral microbial heritage – a mass microbial extinction, if you will. This is seen with immigrants from traditional cultures who lose valuable gut microbes when they move to industrialised nations, correlating with a rise in Westernised diseases in these populations.

High dietary fibre consumption comes with a 30 per cent decrease in death from all causes, particularly some of the top killers globally, including heart disease and type 2 diabetes. Not only that, eating a fibre-rich diet is linked to reduced risk of colon cancer and can reduce inflammation associated with joint pain and arthritis.[13,14,15]

Yet despite now having a clear picture that fibre is good for us, there are still some misunderstandings around how much we need. A good rule of thumb for the biggest impact on health is to aim for *over 30 different plant foods per week*,[16] while eating 10 or fewer has a potentially damaging impact on the microbiota. The tricky thing about fibre is that it's not a monolith, but an intrinsic part of many whole plants. There are over 100 different types of fibre found in foods such as fruit, veg, nuts and seeds, legumes and whole grains. Aim to include some resistant starches – so called because they are resistant to digestion in your gut, but not for your gut bugs. These include cooked and cooled white potatoes, oats, lentils and rice. Some specific fibres such as beta-glucans from oats or mushrooms are known to have direct antimicrobial and anti-inflammatory activity. It may not be as sexy as the latest superfood, and we still have much to learn about fibre to be able to use it as a way to prevent and treat disease, but for those of us striving to feel good, simply adding fibre is a good place to start. Increase it slowly to give your gut time to adjust and take care to drink enough water – rapidly increasing your fibre intake can lead to constipation. And a call to all those parents out there: kids don't need as much fibre as adults, but it's still important. Feeding kids can be tricky (trust me, I know), but look at the foods you child is eating. Many fast and convenience foods are fibre-poor. Good sources tend to be found in 'whole plant foods' that are close to their natural form. For example, smoothies and juices are not a great source of fibre compared with the whole fruit. Here is a guide:

Age (years)	Recommended intake of fibre
2–5	15g per day
5–11	20g per day
11–16	25g per day
17 and over	30g per day

Sorry, Low Carbers, Your Microbes Are Just Not that into You

Fibre is a form of carbohydrate. While many people report wonderful results from a low-carb diet, the effect on the gut microbiota is not well understood. A potential challenge of eating a low-carbohydrate diet, if implemented improperly, is the resulting low fibre intake and knock-on impact on your fibre-loving gut microbes.

A mouse microbiome will shrink ten-fold on a low-fibre diet and there are hints that the same is true in humans. When we eat a fibre-deficient diet, our gut microbes starve, with a knock-on negative impact on our immunity that could open us up to illness down the line.[17] This is especially true if the few remaining dietary carbs contain a limited amount of fibrous food for gut bacteria, which means they have less to eat and less to do.

High-fat diets are interesting because the organisms in the gut that love fat are not those organisms believed to be the most beneficial for the microbiome. So while there have been dramatic health effects reported in people on a high-fat diet devoid of complex carbohydrates, the research is just getting started on what this does to the gut microbiota and the long-term impact on our health. But early indications show that unless a complex carbohydrate intake is maintained while eating low carb, exclusive consumption of a high-fat diet is pretty destructive for diversity of the microbiome.

One's Trash, Another's Treasure

So we know fibre is good for the gut bugs that are important to nurture strong immunity. But what is it about fibre that is so good for our health? When your microbes chow down on fibre, they produce a veritable banquet of metabolic trash – by-products known as 'postbiotics'. Consider it your own personalised health pharmacy. These beneficial by-products act at the interface between diet and immunity, changing the personalities of our immune cells, sparking signals that dial down inflammation, reducing all the unpleasant symptoms we experience when sick and switching on our peacekeeping Tregs, activating healing and repairing damage. And this is not just in the gut – postbiotics circulate in the blood, affecting whole-body regulation of the immune system. Hence, gut health can impact whole-body health.

If dietary fibre affects the composition of the microbiota, and the microbiota regulates immunity, then it follows that fibre should have beneficial effects on inflammatory disease. And results in this area are highly promising, with a significant correlation between low dietary fibre intake and elevated blood markers of inflammation, a possible predictor of future chronic disease. Dietary fibre ensures the immune rheostat remains set at a healthy level of responsiveness against infectious bacteria and viruses. In fact, postbiotics from dietary fibre help us fight certain infections. They do so, for example, by turning up the specific virus-fighting cells to put us quickly on the road to recovery when we have a cold or flu. Postbiotic metabolites have even been shown to help align our circadian clock (see p. 128), which plays a major role in governing immunity to infection.

LEAKY GUT – WHAT HAPPENS WHEN BARRIERS BREAK?

Humans have developed a complicated and highly specialised digestive system. Our digestive tract is delicate – a fragile barrier that is just one cell thick. This is where gut health comes in. Your gut works hard to absorb nutrients, while keeping out undigested food, bacteria and potentially harmful things that we've swallowed. At up to 40 square metres, it forms the body's largest interface with the outside world.[18]

This anatomical set-up is deliberate, maximising digestion and absorption of nutrients, while the delicate cells that line our digestive tract keep this barrier tight – like a mucosal 'firewall'. Our microbiota rests atop this barrier, while our immune cells make sure they stay on the right side of it. Our gut is not an impenetrable barrier, though; when we eat a meal, it becomes more 'leaky', helping us to absorb nutrients. In science speak, we call it 'intestinal permeability'. When this happens, our good bacteria and anything else inside our digestive systems can cross out of the gut and trespass the bloodstream (they are only 'good bacteria' while in their rightful place). Up to four hours after a meal, our immune cells all over the body are presented with bits of food and bacteria containing immune-triggering molecular 'danger' barcodes, prompting the immune system to produce a transient inflammation.

A leaky gut is somewhat normal after each meal. In healthy people who have a diverse microbiome and regularly eat a varied and fibre-rich diet, this is short-lived and these irritants don't usually cause anything more than a mild inflammation of a particular part of the gut. And since leaky gut is normal, the body has all sorts of checks and balances in place to ensure that occasional leakiness does not become a problem.

But can it ever be problematic? Proponents of leaky-gut syndrome claim it's the underlying cause of most modern health problems, and there are plenty of reports associating it with several chronic diseases, specifically autoimmune disorders.

There is some evidence that leaky gut plays a role in autoimmune diseases, both of the digestive tract (like coeliac disease) and outside it (including type 1 diabetes and rheumatoid arthritis). A few studies also found increased intestinal permeability in relatives of those with inflammatory bowel disease, who are at an increased risk of developing the disease. Individuals with food allergies often have impaired intestinal barrier function. Despite all these links, however, what we don't know is whether leaky gut is a symptom or a cause. With limited tools available to accurately measure it, you should be wary of treatments offered by people who claim to be able to 'cure leaky gut syndrome'. None has been tested in randomised clinical trials, and they may do more harm than good.

Protecting the barrier

Bacteria in your gut microbiome play an important role as gatekeepers. A balanced and well-fed (think fibre), healthy microbiota reinforces the integrity of the gut, protecting the sanctity of the body. Other dietary factors can impact on gut leakiness: eating a very large meal or one heavy in excessive fat (particularly saturated) or fructose (fruit sugar, in the absence of adequate fibre) are known factors. It's also wise to avoid snacking, grazing or consistently overeating. Hyperglycaemia – prolonged high blood sugar – featuring in poorly controlled diabetes and high cholesterol can compromise the gut barrier too. Although leaky gut is unlikely to be the sole cause of a particular disease, it may well be an exacerbating factor, particularly in Western society, since it's common to eat meals heavy in fat or sugar (aggravating the whole process) and poor in fibre, which is necessary to seal the gut up tight.

Specific nutrients, including vitamins A and D and zinc, are super-important for the gut barrier. Some amino acids such as glutamine, a building block of protein present within bone broth, may also be helpful for digestion, and supplementation has shown some promise to help heal the intestinal barrier in human and animal models. For people with inflammatory gut conditions, getting additional amino acids into their diets may help with some symptoms. Certain drugs (for example aspirin or NSAIDs, antibiotics, the drugs used in chemotherapy and even high doses of vitamin C) are known causes of leaky gut. Try to control stress levels and alcohol intake. But also be aware that leaky gut can occur through normal processes such as the menstrual cycle in the weeks from ovulation to menstruation.[19] Heavy exercise, particularly in hot climates, and ageing can also be factors.

HAVE I GOT A HEALTHY MICROBIOME?

The phrase 'gut health' is the latest nutritional on-trend topic, linked to a plethora of conditions from heart disease and cancer to autoimmunity and allergy. And the list goes on. Media hype has drawn great attention to the topic, and it is possibly one of the fastest-growing bodies of research in the medical sphere. Microbiologists have known for some time that gut bugs are important for our health, but gut health has not only hit mainstream media, it's now a multidisciplinary medical subject with immunologists also getting a seat at the table as we discover just how important these microbes are for health. As work on the microbiome emerges, it's clear that gut health is definitely not a fad. It's here to stay. And it's probably your best starting point when thinking about caring for your microbial diversity. We are an ecosystem. Diversity is good, fostering creativity and great performance in our bodies, in society, in our world. And now our microbial ecosystem is no different. These germs also teach us that collaboration is the root of resistance and resilience when it comes to our health.[20] Diversity protects us. With the focus firmly on diversity, we are only now unravelling the interplay between how *what we eat* and *how we live* influences our own personalised microbial fingerprint for health.

But what does 'gut health' really mean? And has this phrase actually got credibility when it comes to building strong immunity? Part of gut health is taking care of the delicate food-to-poop tube we call the gut – that includes everything from digestion to microbiota. On the one hand, gut health is felt on a daily basis by all of us, as it is one of the few areas where there is a very tight feedback loop, so everything from stool consistency and intestinal transit time, to bloating, indigestion, flatulence or acid reflux – these gut symptoms really have an impact. And they can punch above their weight in terms of affecting quality of life. But most of them are normal to some degree. It's when they persist that we need to see a health professional.

There are five major criteria for a healthy digestive system:

1. Is your digestion working properly?
2. Do you have any specific gut illnesses?
3. Is your microbiota stable and diverse?
4. Is your gut barrier strong?
5. Do you have a good status of physical and mental wellbeing?[21]

Dysbiosis – the term for disturbances in the composition of the microbiota, potentially influencing ill health – refers to an imbalance of this relationship and is broadly due to a bloom of bad bacteria. Dysbiosis can also occur when we lose friendly bugs or lack diversity of them. You may have decided that, based on your history, you need a little extra help with your old friends. Or you have a chronic disease and have heard that gut health might help. Paying attention to your delicate digestive tract certainly can't hurt. And if you do suffer with digestive problems or have health concerns, it may be all the more important. For all of human existence until very recently, the only levers that we had to impact health were lifestyle factors like diet and exercise and cessation of negative factors like smoking or taking drugs. Learning about gut health offers another opportunity to support our health. From medicine and hygiene to diet and lifestyle, we need to begin to consider the impact of our daily choices on our whole selves, and that includes the microscopic bits we can't see.

For me, gut health is the missing link tying conventional and unconventional medicine together. This is a subject pre-dated by ancient intuitions that align with modern science. We even have sayings that describe how the gut can affect us: we often use our 'gut feeling' to make difficult decisions, and when we are nervous about a job interview or a big examination we have 'butterflies' in our stomachs and may need to make a sudden dash to the bathroom. You might have experienced a doctor discounting your anecdotal story. But my hunch is that the more we understand our microbiome, the closer we will get to having a scientific readout for the anecdote because no two people are the same.

ORAL HEALTH IS THE NEW GUT HEALTH

If you think of your digestive tract as a river – an entryway to the gut – the mouth becomes a bacterial community resource.

Every time you swallow, you ingest thousands of bacteria – some bad, but, most importantly, some good. In fact, just like the gut microbiome, there are good bacteria in your mouth that aid the health of your teeth and the rest of your body.

When you are healthy with a good balance of bugs, your oral microbiome speaks with your immune system just like the bugs in your gut. Poor oral health can be a snapshot of your gut microbiome's health and might even be telling of what else is going on. Studies show a clear link between oral and systemic disease. A surfeit of bad bacteria in the mouth has been seen to spill over into inflammatory conditions like rheumatoid arthritis, inflammatory bowel disease and even cardiovascular disease. That means when we look to oral care first, we're quite literally halting disease in its track. Diet is just as important for oral health. Sugary drinks causing tooth decay might be creating problems by reducing the diversity of our oral microbiome. The worst offenders are chlorhexidine-containing mouthwashes, which not only reduce the diversity in the mouth, but are a precipitating factor in many of our modern-day health concerns. If you want to help your oral health, use a mouthwash containing silver nano particles or a weak swirl of hydrogen peroxide.

GUT HEALTH DEPENDS ON MORE THAN WHAT WE EAT

A changing microbiota isn't just the result of our diets but also of our oversanitised environments, increased antibiotic use and modern-day urban lifestyles. These have all wrought havoc on our microbiota. Comparison of faecal samples from Amazonian hunter-gatherers and Andean farmers with industrialised Western populations tells us that Western microbiomes stand out for having less diverse critters.

If we start to think about this in terms of our relationship with our microbiota, it's taken a very long time for these symbiotic relationships to form – yet vital species can be wiped out in a flash. Microbiota insufficiency syndrome is now used to describe the loss of microbiota and their associated functions that were part of our evolutionary past but are now lacking in our industrialised world. Each generation inherits an increasingly impoverished microbiome, hindering vital early-life immune education. This is now known to have long-term consequences, kindling many a modern-day malaise.

DON'T KILL YOUR FRIENDS

Thanks to a century of germaphobia and hyper-sanitation, industrialisation of our food and environment and overuse of antibiotics, we've inadvertently caused a mass extinction of microbes inside us, contributing to a surge in conditions like autoimmune disease, allergies, autism, asthma, diabetes, obesity and more. In short, modern life is a disaster for our microbiome.

Though many of the factors negatively influencing the microbiome are out of our control (type of birth, geographical location and frequency of antibiotic use, among others), there are many that are not. Plus, I've got some science-based accessible tools and tips for nurturing your good bugs and, ergo, your immunity. Let's explore some of them.

Antibiotics – A Blessing and a Curse

In the early 1900s the top killers were infection, infection, infection. You could die from a cut! Then there were antibiotics. When they first came into widespread use in the 1940s, antibiotics were one of the most powerful tools in medicine.

Arguably, no other class of drugs has had such a profound impact on our health in such a short space of time.

Antibiotics are an antimicrobial substance that treats bacterial infections by directly targeting the microbe, either by killing bacteria or preventing them from multiplying. The beneficial impact that antibiotic control of bacterial pathogens has had on our standard of living is difficult to overstate. At the time they came into use, people assumed that all infectious bacteria could be eradicated by these new drugs. But no one anticipated their damaging effects on our good microbes.

Even a single antibiotic treatment can lead to harmful changes in the composition and diversity of the gut flora,[22] wiping out many vital species and disrupting our ecosystem.

Although a course of antibiotics might cure one specific infection, frequent or long-term courses of antibiotics can leave us *more* vulnerable to infections. In fact, frequent or long-term use can make it more than twice as likely that we will suffer colds and other upper-respiratory tract infections. A study of children who were given antibiotics before the age of two showed that a startling 74 per cent of them were, on average, nearly twice as likely to have developed asthma by the time they were eight. The more courses of antibiotics kids received, the more likely they were to develop asthma, eczema and hay fever. Critical developmental milestones for the microbiota (as well as for the child) occur, in particular, during the first five years of life. In fact, antibiotic exposure in children is a critical piece in our long-term health, associated with increased risk of immune-system diseases such as inflammatory bowel disease, obesity, diabetes, asthma and allergies. There are now even reported cases of antibiotic-induced autoimmunity.

By the age of 18, the average child has received between 10 to 20 courses of antibiotics. These drugs have saved countless lives, and it is important that we don't lose sight of that. But whenever they are used, there is collateral damage. And we are only now fully learning how severe that has been.

Might there someday be solutions that don't rely on antibiotics? The human appendix, projecting off the digestive system, is notorious for its tendency to become inflamed (appendicitis), often resulting in surgical removal. But now we know it plays an important role, acting as a reservoir for beneficial gut bacteria. This humble microbial safe house may help us to recover from serious infections and repopulate after an antibiotic bomb goes off. Another interesting area is the use of specific strains of helpful bacteria to potentially fight off recurrent urinary-tract infections instead of antibiotics, which can make things worse in the long term.

Antibiotics are also potentially a contributing factor to rising obesity rates. Farmers have long known that feeding low levels of antibiotics to their animals helps to beef up their beef. They kill off normal gut bacteria that help to metabolise fats, leading to bigger pigs and cows. Could the same thing be happening to us? Low levels of antibiotics (lower than you would use to treat an infection) lead to fatter mice by up to 40 per cent.[23] When you combine this with a highly palatable diet, this gets further boosts of up to 300 per cent.[24] In humans, obese people with 30 per cent fewer gut microbes tend to put on more weight over time.[25]

It's not just Antibiotics that Send Our Microbiots out of Whack

It's not just antibiotics we have to be careful of. A few commonly used prescribed drugs – including metformin,[26] proton pump inhibitors,[27] antihistamines and non-steroidal anti-inflammatory drugs (NSAIDs)[28] – have recently been associated with changes in gut microbiome composition. Twenty-four per cent of 1,000 marketed drugs inhibit at least one strain of microbiota.[29] We can't say yet whether this is for better or worse, and a shift in gut bugs might also be part of the drugs' beneficial action. But while these pharmaceutical drugs can be a valuable tool in the management of disease, we still don't know many of their long-term health consequences. Even a short course of laxatives can cause inflammation, prompt some gut bacteria to vanish, and cause dysbiosis and an immune response.[30]

MICROBES MATTER: HOW TO MINIMISE ANTIBIOTIC DYSBIOSIS

Take antibiotics and other microbiome-influencing drugs only when your doctor prescribes them as absolutely needed; never self-prescribe them (this can be done in some countries but shouldn't be). If you are given them, always finish the full course you are prescribed.

If you are on antibiotics, or are worried about any medications you are taking influencing your good bugs, try probiotics before, during and after the course (just space them as far as you can during the day from antibiotics, which will also kill any good bacteria you are taking).

Consume fibre-rich foods. If possible, aim to consume 30 different plant foods per week. But remember, when introducing any diet switch there is an adjustment period, so if you are looking to increase fibre, take it slow.

ARE WE TOO CLEAN?

Germaphobia is big business, with a huge market for antibacterial products like hand sanitisers and cleansing gels. But the sharp rise in allergies since the 1980s has coincided with this explosion of personal and home-hygiene products, leaving many scientists scratching their heads and wondering: are we *too* clean?

In the late 1980s, the so-called hygiene hypothesis was based on the observation that allergic diseases were less common in children from larger families in rural areas who had more contact with dirt and animals. But the hygiene hypothesis has largely been overthrown by a newer theory (2003), which suggests that rather than being too clean, the most important change in our environment is the loss of contact with our 'old friends' – the many harmless microbes and dirt, dust and general miasma around us from birth. Science has now shown, beyond a shadow of a doubt, that the microbiota plays a significant role in health.

Since the rise in allergies became more prominent, the mix of microbes and dirt we've lived with, eaten, drunk and breathed in has been steadily changing, thanks to our modern-day germaphobia. Our obsessive wiping down of surfaces is cleansing away the good bugs along with the bad, depriving our immune systems of the many non-threatening microbes that help train them without any harm to us.

Regular cleaning *can* help reduce allergy and asthma triggers in your home (such as mould, pet dander and dust mites). But you may want to pause before you reach for the super-strength antibacterial spray. This puts many of us at a cross-roads: should we continue using things like hand sanitiser to avoid disease, or should we embrace germy hands for the sake of our health? First, consider what your hands have recently touched. If somebody just sneezed and then shook your hand, yes, prioritise hygiene. But remember, the body's microbiome is part of its own natural armour. And one of the problems with these antibacterial, antimicrobial soaps is that you're actually scraping off that armour – the skin's natural defence system. So people who take it too far are actually jeopardising themselves. Also, it turns out that hand-washing dishes is preferable to using a dishwasher when it comes to a balanced microbiome and healthy immune system.[31] Why? Because dishwashers sanitise and disinfect dishes, destroying good microbes in the process, while hand-washing enables some valuable ones to stick around. So squeaky clean isn't always a good thing – good news for those who don't own a dishwasher, although I am not sure I can give mine up just yet!

A little context here is important. Strict cleanliness is vital in some environments, like using hand steriliser in hospitals or if you have any underlying health concerns. But although hand sanitisers are effective at killing bacteria, they should be an adjunct to thorough hand-washing, not a replacement for it. The friction and lathering in thorough handwashing have been scientifically shown to be the single-most effective way to reduce the spread of infection and are pivotal in improving global health. So when you've cut raw chicken, clean and sterilise; when you have a wound on your hand, clean and sterilise; but when your children come in from the garden before eating, cleaning with normal soap is sufficient. Plus, not all hand sanitisers are created equal. To be effective they should be at least 60 per cent alcohol. 'Natural' alcohol-free ones might claim to kill germs but are probably not good enough. Also, they might kill 99 per cent of bacteria, but they won't target every infectious germ, including viruses and parasites. Using the same hand steriliser on a routine basis in the home will damage the normal bacteria that inhabit your skin. For some, the harsh chemicals in many conventional cleaning products may even aggravate or trigger asthma and allergy symptoms. Relaxing hygiene won't reunite us with our 'old friends'. One important thing we can do, though, is to stop merely talking about 'being too clean' and start thinking about how we can safely reconnect with the right kind of dirt.

SHOULD YOU PICK YOUR NOSE (AND OTHER GROSS CHILDHOOD HABITS) IN THE QUEST FOR STRONG IMMUNITY?

Should you pick your nose? Don't laugh – it's a serious question! And, in fact, putting aside the social implications, it might be good for your immunity. OK, so I am being somewhat glib, and the evidence doesn't exactly speak to that. But in the same way that kids put a lot of things in their mouths, nose-picking is a form of environmental sampling.

Our nasal secretions trap germs and other parts of our environment that are in the air, so that they don't enter our lungs. Instead, we swallow most of our nose mucus, even if we don't actually pick our noses and eat it. In the gut, these germs act like a vaccine, teaching our immune system what pathogens are in our environment. One study even concluded that snot can help prevent bad bacteria from sticking to teeth.[32] Nature pushes us to do certain things because they're to our advantage.

Babies have been seen sucking on their fingers *in utero* weeks before birth. But the sight of a child with their fingers constantly in their mouth can drive parents crazy, bringing up fears about germs. Nail-biting worries parents for similar reasons. Yet studies suggest that while these habits in young children may indeed increase exposure to microbes, they may not be all bad,[33] with thumb-suckers and nail-biters being less likely to have positive allergic skin tests later in life. Even assuming the protective effect is due to exposure to microbial organisms, we don't know which ones are beneficial or how they actually influence immune function. But perhaps these results allow us to look at such habits through different eyes, as part of a complex lifelong relationship between children and the environments they sample as they grow, shaping their health and physiology in lasting ways.

URBAN IMMUNITY

As well as getting microbes from our mothers and our diets, we also get them from our environment. In a world of hand sanitisers and wet wipes, we can scarcely imagine the pre-industrial lifestyle that resulted in the daily intake of trillions of helpful organisms. Dirt is good. Disconnection from nature is not. Modern urban life is low on microbial diversity and discourages contact with beneficial environmental microbiomes.[34] This affects our microbiomes and, ergo, our immunity.

Humans have both a physical and psychological need to be in nature. And perhaps an immunological one too. The air we breathe carries bacteria, which together with organisms that come mostly from soil and plants, are deposited in our mouths and airways as we breath and swallow.[35] These are known to have potentially beneficial immunological effects. But compared to green rural environments, the microbiota of modern built environments is not only different but less diverse.[36,37] We know that urbanites are more susceptible to allergies and inflammatory disease, and there is clear evidence that childhood exposure to outdoor microbes is linked to a more robust immune system.[38] Studies of farm children who spent time in animal stables and drank farm milk show drastically lower rates of asthma and allergies throughout their lives than their neighbours who did not.[39]

For those of us not living in the countryside, our immune systems may be missing out on those environmental microbes. But how can we 'rewild' our busy selves, while living in cities? Well, it's surprisingly simple – ensure a regular counterbalance, like spending time in a garden, regularly stepping out into the countryside or park, eating straight from a garden or community allotment. Even just digging your fingers in the soil of a potted plant can improve your mood and nourish your immune system.

Expanding our thinking of the microbiome to encompass natural environments means cultivating a connection with them by breathing, playing and digging in them. The more contact we have with dirt and natural environments, the more we let their microbiomes infiltrate and nurture our own. I was raised on a farm, and while I am definitely not allergy-free, such contact can certainly help. And the good thing is that it is inexpensive, and unlikely to have any negative side effects.

THE DIRT ON EATING WILD

Soil has a microbiome and it's where we used to get most of our probiotics. Acquiring soil and other harmless environmental microbes is called 'horizontal transmission'. This process serves to diversify our own flourishing microbial ecosystems. Back in the day, we grew all our own food and pulled it straight from the ground, often tasting it right there without washing it. Today, we are so far removed from the dirt that grows our food that we're missing out on some of the crucial micro-organisms in our soil that used to populate our own microbiomes.

In 2004 researchers published a paper containing unexpected results. Injecting lung cancer patients with a common, harmless soil bacterium (*Mycobacterium vaccae*, which had previously shown promise for its ability to fight tuberculosis[40,41]) helped patients' quality of life. They were happier and expressed more vitality.[42] Research also suggests that eating trace amounts of this bacterium from garden vegetables, or breathing it in, may significantly help our immunity.

In some cultures, 'geophagy' (eating dirt) is a little more extreme. The most usual time for eating dirt in many societies (the only time in some) is during pregnancy.[43] Most commonly consumed are clays in sub-Saharan Africa. This ancient practice is considered to nurture the mother's immune system via her microbiome. Monkeys that regularly eat dirt have lower parasite loads.[44] And if we look at how kids play when left to their own devices, their impulse is to be on the ground and get dirty with gusto. Eating dirt is particularly common in very young children under two, and before they are old enough for you to talk them out of this behaviour, they normally give it up on their own. But why they do it was largely obscure to all but children – until our recent understanding of the close link between populating a developing microbiome for correct maturation of the immune system.

Now I am not suggesting we all start eating dirt, as there are definitely risks there too. But gentle exposure to germs that live in natural environments by getting outdoors into parks and, where possible, the countryside, particularly during childhood, has lasting benefits for our immunity. It appears that there are mechanisms controlling the inventory of different types of immune cells and that childhood may be the best window of opportunity to set the immune thermostat.[38]

Local farmers' markets are another possible source of microbes. Their delivery vehicles for local produce put dirt back into the diet and, in the process, reacquainted the human immune system with some 'old friends'. Then there is the increasingly popular foraging – nature's free gift of wild food. Again, it gets you outside and is great for the microbiome.

The suggestion that we embrace some 'old friends' does not mean that we are inviting more bad germs too – quite the contrary. Quite simply, what won't kill you – and 99 per cent of the germs surrounding us at any one time in most of the world *won't* – may bring balance to your immune system.

PROBIOTICS – YES, IT'S COMPLICATED, BUT IT DOESN'T NEED TO BE

The principle behind using probiotics is simple: helping our own unique populations of good bacteria to do their jobs. But in practice, with good-gut marketing mania at an all-time high, it's a confusing world, and products advertising live probiotic cultures – from pills to functional foods to probiotic mattresses and pillows – are an unstoppable trend. You might have tried some fermented food or probiotic capsules, but what exactly are probiotics, and how do you know if they actually provide any health benefits?

The definition of probiotics is: 'live micro-organisms that when administered in adequate amounts confer a health benefit on the host'. 'Live' refers to survivability – that means enduring the trip from the manufacturer to your home and then from your mouth to your gut, and that's no easy journey. They must survive the harsh environments of the digestive system (the stomach, after all, is designed to prevent the passage of any bacteria swallowed orally) and land in your colon in the latter part of your digestive tract.

So let go of that kombucha, and let's separate fact from fiction. Scientifically speaking, few products on the shelf will meet the official definition of a probiotic. Only if a beneficial health effect is demonstrated on human subjects in a controlled study can a product maintain, strictly speaking, that it's a probiotic. Because the term 'probiotic' implies a health claim, as of 2019 the European Food Safety Authority demands that health-food companies supply scientific evidence to back up their assertions regarding probiotic food products like yoghurts and drinks in Europe. They are also trying to place a ban on using the word 'probiotic' on food packaging in the absence of clinical evidence so as not to mislead consumers.

This probably still leaves you with a lot of questions, though. So where *can* you find a scientifically validated immune-nourishing probiotic? And how is it best to take it? Even with the expensive gut health tests available online, we are not yet at a stage where we can make predictions about which strains of bacteria you need. We can measure who's there and who's not there – but we can't always tell how meaningful this is to our health. Just like our fingerprints, the individual differences in our microbiome make us unique. And it's not just the strains, but the output of your entire microbe community – that level of individuality is something we just can't measure yet and is a major limitation of most stool-testing kits available commercially. While the diagnostics currently out there are certainly fun to do, they can't give you scientifically substantiated, actionable information. You can ask questions about what is known about the particular bacterial strains these gut health tests find: have they been studied, and in what quantities do they actually do something in your body?

You might also be wondering, if we are all so different, then why should we take the same probiotics? Good point. But for probiotics, the gold standard is what we call 'efficacy in a heterogeneous population'. This means being able to show that an intervention has a beneficial effect irrespective of an individual's starting microbiome.

Probiotics: How They Work – And How They Don't

There is a growing and, frankly, concerning trend towards a one-size-fits-all probiotic recommendation for gut health. Unfortunately, when it comes to probiotics, the overall verdict is not as promising as the adverts might suggest. We still lack detailed knowledge, and, unlike medicines, probiotics are not guaranteed to give the same (or even similar) effects for each person.

There is a common misconception that probiotics must 'colonise' or alter the composition of your microbiome to work. This is not true. In fact, outside of specific cases like faecal transplants, there is little evidence that probiotics colonise, or that they need to. Compared with the tens of trillions of microbes already rooted in your gastrointestinal tract, most ingested from probiotics are more allochthonous members – meaning they do not take up residence within our gut ecosystem, but are mostly rather transient inhabitants, detectable only for a limited time during frequent consumption. This is not a reason to dismiss their health benefits, though. They are known to be biologically active in other ways, including promoting nutrient availability from our food and producing bioactive compounds that strengthen immunity and shift away from anti-inflammation as they pass through our gut.

Probiotics also don't contain enough new bacteria to make a significant difference to your established ecosystem. Even if they did, we don't know enough about the safety of introducing colonising microbes. Large numbers of newcomers moving in and displacing your existing bacteria could alter the unique balance of your ecosystem and trigger unintended consequences. What scientists do know, however, is that, as transient microbes, probiotics travel through your GI tract, interacting with your immune cells, dendritic cells, gut cells, dietary nutrients and existing bacteria to, directly and indirectly, deliver benefits.

Some transient microbes help with a tight gut barrier. Others trigger neurotransmitters that stimulate muscle contractions for better, more regular poops. Still others produce helpful by-products shown to be beneficial for immune health. This is why, if you choose to take a probiotic, continuous daily intake is important as it's unlikely those bugs hang around.

The strongest evidence supporting the use of probiotics is related to the treatment of diarrhoea and digestive upset – after taking antibiotics, for example.[45] But probiotics can be helpful not only for digestive health, but also in supporting proper immunity. We still don't know if they are useful for everyone or which specific strains should be taken, but clinical trials with many strains have been shown to help reduce recurrent infections, improve inflammation and even prevent allergy and eczema in high-risk children.[46] There's also some evidence suggesting that probiotics may promote better ageing by reducing inflammation that typically occurs as you grow older.[47,48] Taking probiotic supplements is also linked to reduced likelihood of getting colds, and making them shorter in duration and less severe when they do strike.[49] Of course, nothing is an immediate preventative or cures symptoms right away. If you feel sick and eat some probiotic yoghurt, you're not necessarily going to feel better. The quality of much of the research in this area tends to be poor, and sometimes inconsistent, probably impacted by the fact we are all so different in our microbiomes to start with – another reason to aim for a food-first fibre approach when looking after your good gut bugs.

If you choose to take a probiotic, you should look for strains that have evidence to back their beneficial effect for your condition. And even then, there is no guarantee that they will work.

How Diet Can Improve Gut Health without Probiotic Supplements

- **Go organic.** Organic produce has a significantly more diverse bacteria population than its conventionally grown counterpart (especially when consumed raw, since cooking would destroy these good bugs).[50]
- **Ferment.** Like your microbiome itself, fermented food and drinks like kimchi, kombucha, kefir and sauerkraut contain a natural synergy of many different types of yeasts and bacteria, as a result of these microbes all living and thriving in a controlled way (unless they are pasteurised as part of processing, which causes the good bugs to die off). These aren't strictly probiotics in the scientific sense, although they may contain probiotic bugs. Anecdotally, I have seen many benefits from consuming fermented foods. But remember, the clinical proof for this is limited with the exception of fermented dairy, which has been shown to benefit overall health, even though the evidence is very, very thin.[51]
- **Aim for 30 plus.** While we know that people who eat a greater number of different plant-based foods have more diverse, healthier microbiomes, less is known about exactly which bacteria prefer which foods. It's particularly important to get fibre into kids when they start eating solid food to ensure that you are cultivating a diverse gut ecosystem right from the beginning.
- **Be label savvy.** Certain dietary patterns high in salt, sugar and fat can negatively impact our microbiomes. Find out more about the negative impacts of these foods in Chapter 7.

CHAPTER 4

Sleep, Seasonality and Circadian Rhythms

'The shorter you sleep, the shorter your life.'
MATTHEW WALKER, Professor of Neuroscience and Psychology, University of California, Berkeley and author of *Why We Sleep*

I generally maintain quite good health (as one would hope in my line of work), with the occasional slight cold, ache or pain. Last winter I ended up quite unwell, though. It started with a seasonal January cough and cold, a scratch at the back of the throat shared with my husband and son. Within a week, they were both well on their way to recovery and I assumed I wouldn't be far behind. I was wrong. But not wanting to let down colleagues, family or friends, I dismissed the lurgy and carried on. It was a particularly busy time and I had (as I often do) overcommitted. At night I lay awake while my brain ticked over, then felt stressed each morning, trying to figure out how I was going to keep on top of my to-do list. Still, so convinced was I that these germs would soon be gone, I powered on as normal.

Until one day, two months later, I couldn't get out of bed. I couldn't make my kids' breakfast or take them to nursery. No more business as usual. I spent three weeks in bed with pneumonia, with death-rattle coughing fits, hacking up 'green goo' and my husband having to help me turn over in bed because my ribs hurt so much. I'll spare you more details. But why am I telling you this? Because the reason a little seasonal cold developed into pneumonia was not because of a lack of immune-'boosting' superfoods or even a poor diet. It was because I had neglected arguably the most important pillar of health – the foundation upon which all the others sit: sleep. My sleep had been eroded by stress. My doctor called it 'modern-day Mummy burnout': juggling work life with family life.

I was embarrassed not to be bulletproof, but it may have been the best thing that ever happened to me – because it forced me to take action. I have now completely changed how I view sleep and stress and their close relationship. There is no part of our physical or mental health that is untouched by poor sleep. It is intimately entwined with immunity, yet neglected in much of the wellness sphere, which favours supplements and functional foods. It's a hard pill to swallow – knowing there is no pill to swallow.

One of our least valuable clichés is that you can sleep when you die. Much truer is: if you don't sleep, you will die – sooner. But in today's world, good sleep can be hard to come by. In our technology-rich environment, there is always something distracting us from just going to bed on time. Sleep requires discipline and habit. But as you will read in this chapter, the investment is worth it.

HOW MUCH SLEEP DO WE NEED?

It's clear that we all have a basal sleep need and require a certain amount to feel healthy (and sane). We know it's impossible to do without it completely – just think of what happens when you try to pull more than one all-nighter (intentionally or not). But exactly how much sleep do we need?

Some good signs of getting enough sleep would be waking up without an alarm clock, not needing caffeine to keep you going and being alert throughout the day. The day after a sleepless night, we feel irritable, exhausted, unhappy and stressed. If you, like me, are a parent of small children and have a full-time job, you'll know how precious even a few hours of sleep can be in retaining your sense of sanity. Most of us need, on average, around seven to nine hours of sleep each night, and moving below this by even a small amount poses a risk of early death.[1] If you have a heavy life load, are particularly stressed or busy or training hard at the gym (or you're an athlete), you should be shooting more towards the top end of that figure because sleep is our recovery time. And it's not just about quantity. Quality matters too. And this can be much harder for us to gauge.

The definition of good-quality sleep is being asleep for more than 85 per cent of the time you're in bed, waking up no more than once per night and for fewer than 20 minutes. So if you are in bed for 10 hours, 8.5 of them should be spent sleeping. Given that most of us don't fall asleep immediately when we get into bed, it can be challenging to fit in a healthy number of sleeping hours. People in the UK are sleeping less and less, with 77 per cent not reaching their sleep needs. Plus, people experiencing inadequate sleep are also more likely to suffer from chronic diseases such as hypertension, diabetes, depression and obesity. There is even the classification of a new sleep disorder – insufficient sleep syndrome – whereby people actively choose to get less sleep through prioritising certain lifestyle habits (Netflix and sleep don't always mix) that impact on their health. We are running a huge experiment with our biology, and the early outcomes are not looking favourable.

SLEEP ARCHITECTURE

Sleep is made up of stages, collectively known as sleep architecture. The stages comprise two types of sleep: rapid eye movement (REM) sleep and the progressively deeper non-rapid eye movement (NREM) sleep.

Each stage can only be reached by passing through the previous one:

- **Stage 1 NREM:** light sleep – your eyes are closed, but it's easy to wake you up.
- **Stage 2 NREM:** as your body is getting ready for deep sleep, your heart rate slows and your body temperature drops. This stage takes up about 50 per cent of your time across the entire night. Think of it as the stock of sleep soup because not only does it provide the platform upon which the remaining stages sit, but it's pretty nutritious all by itself.
- **Stage 3 NREM:** known as slow-wave sleep, this is when you are deeply asleep. It's harder to rouse you during this stage, and if someone wakes you up, you'll feel disoriented for a few minutes. Most deep sleep tends to happen early in the first half of the night, which is why going to bed early can be helpful.
- **REM sleep:** this is when you dream, usually 90 minutes after you fall asleep. The first period of REM typically lasts 10 minutes. Each of your later REM stages gets longer, and the final one may last up to an hour. Your heart rate and breathing quicken. You can have intense dreams during REM sleep, since your brain is more active. Babies can spend up to 50 per cent of their sleep in the REM stage, compared with only about 20 per cent for adults.

All stages of sleep are important, but they each provide something different for our bodies. During the deep stages of NREM sleep, the body repairs and regrows. This is our immunity-fortification phase in preparation for the challenges of each day.

REST IS THE BEST

Sleep and immunity are bidirectionally linked. Immunity alters sleep, and sleep can impact our bodies' defence systems. While more sleep won't necessarily make you invincible, lack of sleep almost immediately tips your immune system into imbalance, simultaneously dampening parts of it and empowering others.

Now that might sound contradictory. But of course, immunity is not binary, nor is it straightforward. Sleeplessness turns the once balanced system into reckless pinballs of powerful inflammatory molecules bouncing off your body's bumper rails, and sometimes through them. Studies show time and time again that lack of sleep profoundly compromises both immunity and recovery.

An early indication of just how important sleep is for our defence system came in the late 19th century, when studies showed that total sleep deprivation in dogs led to death after several days due entirely to a breakdown in immunity.[2] In fact, we now know that a single night of poor sleep leads to a dramatic decrease in natural killer cells – our first-line defence against viruses and potentially cancerous cells. This decrease – up to a 70 per cent reduction – is not to be sniffed at, lowering surveillance of our body and creating similar negative changes seen in other immune cell types.[3] If you are always struggling with niggling colds throughout the year, check you are getting enough sleep of a decent quality.[4] People who sleep six hours a night or fewer are four times more likely to catch a cold when exposed to the virus, compared with those who spend more than seven hours a night asleep. This alone may be the reason you end up with a bad case of the flu after the Christmas party season. Plus, research also shows that poor sleep causes vaccines to be less effective, which is problematic for more serious illnesses and in vulnerable populations such as the elderly.

But it's not just about whether or not we come down with a cold or flu. Sleep also influences how we fight illnesses once we have them. When our immunity is challenged, say by an infection, and inflammation is triggered, it will, depending on its magnitude and duration, do several things to our sleep: it makes us tired, increasing the desire to sleep; it also disrupts the quality of our sleep, so we may be spending more time asleep, but not getting that deep, restorative type.

We can all recall times when we were under the weather but tried to fight that feeling that was telling us to rest up while the body fights off the infection. This is an example of the immune system using inflammatory cytokines to directly communicate to the brain to change our behaviour. This immune-induced tiredness and sleepiness encourages us to be less active. Because while we sleep, we get a better immune response (perhaps the only way to truly boost our immunity) to the infecting germs, facilitating a faster recovery. This is why fevers tend to rise at night. But if we are not sleeping, our fever reaction is not primed, so we may not be waging war on infection in the best way we can.

Even when we are well, a good night's sleep really is the best medicine. Each night of sleep elicits important changes, fortifying our entire immune system. It keeps bone-marrow stem cells young – these are the 'blank canvas' precursor cells that slowly regenerate your entire immune system by developing into fresh new immune cells every day. Conversely, sleep deprivation leads to 'sleepy' stem cells in our bone marrow that are functionally impaired due to genetic changes that stop them from leaving and entering the blood where they replenish our circulating pool. Luckily, these genetic changes are corrected by a good night's sleep.

Deep sleep is particularly important for transforming our fragile, recently formed memories into stable, long-term ones. Now, we also know that deep sleep also strengthens immunological memories of previously encountered germs and dangers.[5]

A LIFE-OR-DEATH ISSUE

What is truly worrying is that our sleep (or lack thereof) plays a key role in our ability to fight off serious and chronic modern-day health conditions. The rise in unruly 'pro-inflammatory' signals when you don't sleep enough[6] can not only exacerbate any underlying conditions – like allergies and autoimmunity – but may even increase our risk.[7,8] Plus, if we don't get enough deep sleep, our capacity to deal with pain is drastically reduced. We are more likely to suffer from aches, headaches or a worsening of any underlying conditions.[9]

We know that insomnia is a predictor for heart disease and metabolic complications due to the erosion of our delicate immune balance. Short sleep, difficulty falling asleep and difficulty staying asleep increase type 2 diabetes risk by 28 per cent, 57 per cent and 84 per cent respectively. This may be because disturbed sleep alters our metabolism in a way that means it is less effective at processing our food. Reducing healthy people's slow-wave sleep for just three nights manifests itself in a remarkable way: a change from potentially good blood-sugar management, to something akin to a rapid onset of pre-diabetes and increasing risk of obesity in kids by 45 per cent. Poor sleep quality may speed up ageing of gene protectors, which can increase the risk of cancer. In fact, people who sleep less are actually more likely than their well-rested counterparts to die from all causes. Sleep would appear to be anything but expendable.

SLEEP AND THE SELF-CLEANING BRAIN

Sleep, it turns out, may play a crucial role in our brain's physical and physiological maintenance. As our bodies sleep, our brains quite actively play the part of mental janitor, clearing out the junk that has built up as a result of our daily thinking.

The lymphatic system is a vascular one – much like our blood circulatory system but with key differences that you will meet formally in Chapter 6. It is the superhighway that immune cells use to whizz around the body. It also serves as the body's custodian; whenever waste is formed, it sweeps it clean. Glymphatics – the very recently discovered special lymphatics of the brain – switch on, while we are sleeping, working hard to remove all our brain's metabolic waste from the day. But not just any sleep will do; deep, slow-wave NREM sleep is particularly important for this special system to function. During deep, slow-wave sleep, the glymphatic system clears as much as 40 per cent of the total amyloid-beta build-up (a protein seen accumulating in the brains of Alzheimer's sufferers). A mere 36 hours of sleep deprivation increase these amyloid levels by 25 to 30 per cent. So when we skip sleep in our youth, we may be doing irreparable damage to the brain, prematurely ageing it or setting it up for heightened vulnerability to other insults later on. Modern society is increasingly ill equipped to provide our brains with the requisite cleaning time.

Beyond the brain, impairing the vital glymphatic function is worryingly emerging as a link to high blood pressure, cardiovascular disease and even diabetes. At the (relatively) more benign end – after an all-nighter or the extra-stressful week when we catch only a few hours a night – sleep deprivation impedes our ability to concentrate and analyse information creatively. At the extreme end, chronic lack of quality sleep from shift work, insomnia and the like could accelerate neurodegenerative disorders.

THE TRICKLE-DOWN IMPACT OF SLEEPLESSNESS

Too little or disturbed-quality sleep not only changes our immunity, leaving us open to illness, it changes our behaviour and emotional state. We become grumpy and more likely to struggle with social withdrawal and loneliness.[10] Sleep loss is linked to an increased likelihood of partaking in poor lifestyle choices – food preferences being one.

Brain scans in sleep-deprived subjects show increased activity in reward pathways that make high-energy food more appealing. Obesity and sleep exist in a vicious cycle: staying up late increases calorie consumption but it's normally at a time when we are lying around not using up much energy. Short sleepers have a 45 per cent increased chance of developing obesity. And being obese, in turn, is a known sleep disrupter.

SOLAR POWERED

Sleep is regulated by two main processes. First is something called sleep–wake homeostasis – a feedback mechanism controlled in the brain that governs how much you need. It is quite intuitive in its operation and can be thought of as a kind of internal timer, generating a pressure for sleep as a function of the amount of time elapsed since waking. Basically, the longer we have been awake, the more likely that we will feel like sleeping. The best known of these sleep-regulating substances (although probably not the only one) is a build-up of something called adenosine in the brain (caffeine works by blocking the function of adenosine). Through this feedback mechanism, sleep generally increases in duration and intensity after a prolonged sleepless period.

As much as we all sleep, you may also realise that in addition we (most of the time) follow a daily rhythm of *when* we sleep. This is generally independent of how long we have been awake, and is the second process regulating sleep, known as the circadian rhythm. Our circadian rhythm is an (almost) 24-hour internal clock running in the background of our bodies. It is responsible for imposing and synchronising a 24-hour rhythm on our behaviour and body functions, including the propensity to sleep or be awake. We all run on our own roughly 24-hour circadian rhythm, but not every human clock runs the same. And so circadian rhythms are independent of how much prior sleep we have had.

If you imagine staying up late at a party, you might start to feel tired around your normal bedtime – this is influenced by your circadian rhythm. But if you push past that and stay up way later than normal, you might find that you feel quite awake and alert. This difference in alertness (pushing past circadian control of sleepiness) is separate from the sleep pressure that builds up the longer we have been awake.

A master circadian clock in the brain dictates the time of day to the rest of our body systems. Where the nerve fibres from our eyes criss-cross and enter our brain, there is a small cluster of only 20,000 neurons called the suprachiasmatic nucleus (SCN). It's important that our master circadian clock in the brain has a dedicated link to the eye. This ensures that outside information concerning sunlight and darkness is directly transferred from the eye to the master clock. Humans are sensitive to light with a short wavelength – specifically, a blue light with a particular wavelength emitted from the sun. Once the master clock is in sync with the outside day–night cycle, it can communicate this information to the rest of the body: a rhythmic feedback loop. Production of the hormone melatonin (responsible for making us feel sleepy) is a key example. As darkness sets in and blue light sets with the sun, our eyes relay this information to the brain, signalling for melatonin to be released. For the next few hours, as it becomes dark, more melatonin is secreted, which signals the brain to go into sleep mode. When the sun rises, melatonin secretion is inhibited, and the brain's awake circuits resume. If we are exposed to blue light in the evening, sleepy melatonin is reduced, leaving us feeling wakeful, influencing how long it takes to get to sleep and also disrupting sleep quality.

Melatonin is a powerful tool in our health arsenal. It turns on DNA repair and antioxidants, it's protective in the ageing process and plays an important role in immune regulation. Other benefits of nightly melatonin production are regular menstrual cycles, enhanced mood, improved brain health and cancer-fighting properties. The other reason melatonin is extremely important for our health has to do with the link between melatonin and serotonin. Serotonin is a known mood-enhancer, and it's also the precursor for melatonin, meaning that melatonin is made from it. When there is enough of both of these in our brains, we rest better and feel happier. A lot of people experience sleeping problems and associated poor mental health issues when their brains don't produce the right amount of melatonin, or the circadian rhythms shift and it isn't made at the right time.

Our circadian clock is not a rigid timing system. Rather, it must be synchronised daily to coincide with the rotational cycle of the earth. This alignment is defined as entrainment.

Without entraining signals, known as zeitgebers, our clocks would run slightly off of 24 hours, eventually misaligning with the natural rhythm of day and night. Light is also indirectly responsible for the timing of food intake, another powerful entrainer of rhythm. This means that *when* we eat, irrespective of what we eat, may also be the source of a good night's sleep. Eating within a 12-hour window – as our great-grandparents did – is shaping up to be a useful tool in staving off markers of poor health. Dampened circadian rhythms also occur when we are given unlimited access to a high-calorie diet. Collectively, our peripheral body clock senses mealtimes and other non-light zeitgebers, including movement, social contact, changes in temperature and sound, integrating this information with the sensing of daylight and darkness by our eyes, to keep our circadian rhythms on time.

CIRCADIAN IMMUNITY

From bacteria to mammals, nearly all organisms have adapted their physiology and behaviour to follow a daily rhythm. This, for the majority of species, is dictated primarily by light and dark, day and night. We humans have adapted to sleep during darkness and are at our most active during daylight. This pattern has influenced virtually all aspects of our physiological architecture, from the expression of our genes to our behaviour, making it essential to our survival before we had 24/7 lighting.

Our body's daily circadian rhythm has affected our well-being in all kinds of ways. Night is biologically designated sleep time. Our alertness changes, but so do our various hormones, digestive processes, exercise capacity, blood-sugar control, as well as our immunity. Rhythmic circadian genetic switches are present in the majority of our immune cells. It is not surprising, then, that several features of the immune response are profoundly regulated in a time-of-day-dependent manner. Rather than a peak time for strong immunity, it's more a case of our immunity being in a different state, depending on the time of day. These oscillating ebbs and flows have been crafted by evolution and are of no surprise when we consider the daily rhythm of life. But it's the regular rhythm-keeping that's important. This gives our elegant defence a precise, integrated and cohesive response across a 24-hour period, conserving energy: the immune system only needs to prioritise being 'poised for attack' as we enter our active daytime period, then resolution and repair can happen while we rest. The reason for these fluctuations may be an evolutionary adaptation to deal with germs that attack us at different hours, or perhaps because it has no choice: less a specific benefit and more a side effect of our daily 24-hour cycle. Circadian peaks and troughs of immunity can certainly change how susceptible we are to infection or how likely we are to experience a flare of an inflammatory condition. But it's the regular rhythm keeping that's important.

Outcomes from surgery, infections and even vaccines can vary depending on the time of day. Heart attacks in humans are known to strike most commonly in the morning, and research suggests that morning heart attacks tend to be more severe. Even our microbiota shows circadian variations, influencing both how we respond to infections and our sleep quality.[11,12] Asthma and autoimmunity exacerbations are more common at night or the early hours of the morning, when inflammation in our blood peaks naturally, adding to an already inflamed state. For rheumatoid arthritis patients, the peak of inflammation is shifted forward to waking in the morning and can be 10-fold higher than in healthy people. Wound healing, which can take days, uses a circadian clock to optimise different aspects of the process at different times of the day or night.[13] But there is no simple rule. Each condition has its own rhythm.

Regardless, messing with our biorhythms definitely has negative health consequences. Alterations to our sleep/wake cycle affect the number of circulating master controller T cells, antibodies, and even those virus-fighting, cancer-seeking NK cells, as well as sending inflammation haywire. When our immunity is challenged during circadian disruption we see impaired immune function and unchecked inflammation. Jetlag causes more than just a feeling of being discombobulated – it makes you more susceptible to getting sick. In fact, even a single shift in our circadian rhythm, without sleep loss, can put us in a worse position when it comes to health.

The prevalence of shift work is relatively high around the globe; approximately 20 per cent of workers in Europe are engaged in it. Shift workers are more likely to suffer from fatigue, sleep deprivation and sleeping disorders like insomnia or sleep apnoea, as well as other health problems, particularly metabolic syndrome[14] and heart disease. There is also a strong link between circadian disruption and cancer risk,[15] albeit other factors such as poor diet and lack of exercise may be at play. In 2007, the World Health Organization classified night-shift work as a probable carcinogen due to circadian disruption. This is the unfortunate reality for those in occupations requiring shift work, and we need to understand how best to protect them.

THE CORTISOL-AWAKENING RESPONSE

You will properly meet cortisol in the next chapter, but for now you only need to know that it is the body's main stress hormone, and it works with different parts of the brain to control mood, fear and motivation. It is primarily associated with 'fight-or-flight' instincts, but also plays a vital role in a number of processes in your body. Unlike melatonin, cortisol starts secreting in the morning when you need to wake up, rising during the day to keep you energised, before falling again in the evening when melatonin secretion begins.

Despite its notoriety, cortisol is the hormone that helps give us the get-up-and-go we need to get out of bed and prepare to face the demands of the day. The cortisol-awakening response (CAR) is the increase in cortisol concentration in the blood that occurs in the first hour after waking. It's the body's way of 'naturally caffeinating'. Our morning CAR doesn't just help us wake up feeling ready to take on the world, it is hugely influential on our overall health. The CAR is a mini stress test, another facet of our circadian rhythm that informs our wellbeing. When we are getting close to waking in the morning, our melatonin starts falling to barely detectable daytime levels. The moment we open our eyes and light comes in, this morning CAR shoots up, reaching its peak after around 30 minutes, before coming back down again. If your CAR falls flat in the morning, it can cause those classic symptoms of feeling run down or just a bit 'off', less able to cope with the demands of the day.

Probably one of the newest areas where CAR is important is autoimmunity. When your body makes immune cells, it first checks them to make sure they aren't autoimmune to you already in your thymus gland (see p. 82). The thymus churns out a huge number of T cells, each with a unique receptor, the aim being to provide you with the broadest possible infection protection. The process is pretty random, so there is always a small possibility that autoimmune cells are generated. The thymus quality checks each cell, and if it does detect an autoimmune reactive T cell, then it gets pulled to the side and tagged for destruction. Cortisol enacts this destruction, called the glucocorticoid-induced apoptosis of the thymus gland. If your CAR flatlines, then those cells aren't necessarily destroyed and instead leak out into circulation to cause or worsen autoimmunity! You might have worse symptoms if you already have a condition like an autoimmune disease.[16]

HOW TO SLEEP WELL IN A MODERN WORLD

Modern life is not conducive to good sleep; and sleeping well (and sleeping right) in today's world takes work. One way to safeguard our sleep is by paying attention to your natural circadian rhythm, but you also need to ensure the best environment and state of mind in which to drift off. Healthy sleep habits, often referred to as good sleep hygiene, can make a big difference to your quality of life. I am going to show you how, from the moment you open your eyes, you can make changes that protect sleep, despite modern-day demands. And I'll explain how implementing simple hacks and rituals that top and tail your day is one of the most underrated ways to nourish your immunity.

Block Out the Blues

Blue light is not just emitted from the sun as it shines. Man-made sources are everywhere, from fluorescent and LED lighting and flat-screen televisions to the display screens of computers, electronic notebooks, smartphones and other digital devices.

Nowadays, 75 per cent of the world's population is exposed to light during the night.[17] Evening exposure to alertness-inducing blue light is messing up our rhythms – we misalign, our clocks get confused, and we feel tired or awake at all the wrong times. Frequent screen users are much more likely to report falling asleep later,[18] sleeping less and waking during the night.[19] It's particularly a big issue among young people, with more than 90 per cent of adolescents' faces buried in screens before bed at a cost of not only poorer academic performance, but also increased risk of health issues such as diabetes and heart disease in later life.[20] It's also worth noting that young kids (under the age of 10)[21,22] are much more sensitive to the melatonin-suppressing effects of blue light in the evening. We can also pass on the damaging effects of night-time light exposure to our unborn children through epigenetic changes to their DNA, adding to a growing body of evidence that there's a health cost to our increasingly illuminated nights.[23]

Luckily, limiting screen use for a week can restore normal sleep patterns.[24] Seems like an easy fix, but as more and more of us need or want to use smartphones, tablets and computers during evening hours, this may not be a solution for everyone. In fact, writing this book, I am staring at a screen into the wee small hours of the night as I struggle to fit it in around a full-time job and two small kids.

Blue-light-blocking glasses

Wearing good-quality blue-light-blocking glasses is almost as effective as no screens at all.[25,26] These are relatively cheap and becoming more widely available online. There are also applications on phones and laptops that shift the blueness of light emitted with the advancing time of day. Even so, I'd still try to reserve the last hour or so before bed for not doing anything too stimulating. Scrolling social media can be draining and anxiety-causing. Putting your phone down a little while before bed is a good habit to get into, not just for blocking out blue light, but also as a way to unwind and invite that sleepy feeling in. Also, keep evening light low by using red incandescent light bulbs in lamps.

Wake Up, Get Lit

As well as getting too much blue light at night, many of us don't get enough natural sunlight during the daytime. In fact, our total light exposure during the day may be more important in regulating our sleep. People who are exposed to light before midday, say, by walking to work in the morning, have better sleep quality. Natural light when we wake up may even help undo some of the melatonin-dampening effects of using electronic devices before bed.[27] Getting plenty of sunshine in the daytime is also important for maintaining healthy levels of vitamin D. Research links vitamin D levels to sleep quality. In fact, several studies associate low levels of vitamin D in the blood to a higher risk of sleep disturbances, poorer sleep quality and reduced sleep duration.[28,29]

Bank Some Sleep

Aiming for seven to nine hours' quality sleep per night matters, but as schedules can get busy and, at times, erratic, a more helpful strategy is to consider these sleep hours across the week, rather than each night.

There may be days when you tried to get a good night's sleep, but something got in the way. Trying to make up for that lost sleep across the week, incorporating naps as well (see p. 139), will help to erase some of that sleep debt. For example, eight hours per night for a week is 56 hours. But if on one night, you only manage six, you can try to compensate another night to get what you need across the whole week. You can also do that in reverse. If you know you have a late night coming up or will be sleeping in a different place, so may not get the best-quality sleep, you can bank some sleep ahead of time.

Although we know that vital maintenance of our immunity happens during deep NREM slow-wave sleep, it's quite difficult to control the ratio of your different sleep stages. There is a little bit of research showing, for example, that going to bed early, increasing fibre intake and regular exercise will help you get that slow-wave sleep. But in the grand scheme of things, it's a very, very small effect.

Keep It Cool

What you might not have considered is that it's beneficial to keep your bedroom cold year-round for optimal sleep. As night-time approaches, your body temperature bottoms out right before bed, signalling that it's time to slow down and get some rest, and helping to trigger sleepy melatonin. By 4.30 am the body is at its coldest and the temperature starts to rise naturally as you get closer to waking. By keeping your bedroom cooler (16–18°C), you're reinforcing your body's natural instinct to sleep.

Ironically, a hot bath before bed can help cool you down. As soon as you step out of the bath or shower, your body temperature drops rapidly to re-regulate with the temperature of the room. That quick change physiologically can help encourage a drop in your core temperature that causes sleepiness.

And it may seem counterintuitive, but drinking a hot herbal tea can be useful too. Hot drinks increase the body's heat load and it responds to that by sweating. The output of sweat is greater than the internal heat gain, and this is where it all starts to make sense – when the sweat evaporates from the skin, it cools you down.

Map Out Your Mealtimes

I mentioned earlier that mealtimes are also important zeitgebers in setting your circadian rhythm. Insulin, a hormone released when we eat, adjusts circadian rhythms in many different cells and tissues individually by stimulating the production of a protein called PERIOD, an essential cog within every cell's circadian clock. But many of us spend more than 18 hours a day snacking instead of eating at regular mealtimes. Best suggestions are that tucking into your meals at regular intervals within a 10–12-hour window could help you get a good night's sleep.

Daytime Movement

Studies indicate that sleep may receive some of its most significant benefits from exercise that is consistent and routine over time, especially for people who experience difficulty sleeping. In particular, aerobic exercise – any activity that gets your blood pumping and large muscle groups working – is an effective strategy to improve sleep quality and duration, as well as daytime wakefulness and vitality. It doesn't have to mean going to the gym – just very short bursts of activity regularly throughout the day. Examples of aerobic exercise include brisk walking or swimming. Morning exercise, especially, gives a particular boost to deep sleep. Just be aware that vigorous exercising too close to bedtime may make it more difficult to sleep because of rises in your core temperature. Aim to finish your workout at least two hours before bedtime.

Rhythm and Routine

As much as we may believe we can undermine our own biology, our bodies are just not designed to be awake and active during the night. People who have erratic sleep schedules due to overnight, changing and unpredictable work hours may notice the toll it takes on their health. These people are possibly the most vulnerable to sleep-dysregulated ill health, so we need to have strategies to protect them. It is very challenging, but best efforts include trying to play with light exposure and meal timings. Trying not to eat large meals or snacks continually during your shift is really important too, due to the metabolic derangement that takes place. And once you do get off [illegible] shift, make sure your sleeping environment is what it [illegible] at night, using blackout shades and eye masks, and [illegible] hat there is not a lot of noise. When trying to revert back to a day shift, be sure to get enough morning light before lunchtime.

Use Tech

Although you should try to limit screen tech in the evening, not all devices should be avoided. These days there are a bunch of sleep devices that may help you sleep – from smart mattresses to white-noise machines and apps that help you track your sleep and give you actionable feedback. Just be aware that if you suffer quite badly from insomnia or feeling stressed out, tracking devices might not be the best idea. 'Orthoinsomnia' – an excessive use of sleep tracking and sleep anxiety – is a legit condition, resulting in sleeplessness from a perfectionist quest to achieve ideal sleep data through obsessively overusing sleep-tracking devices.

Power Naps

More than 85 per cent of mammalian species are polyphasic sleepers, meaning that they sleep for short periods throughout the day. Humans are part of the minority of monophasic sleepers, meaning that our days are divided into two distinct periods: sleep and wakefulness. It is not clear, however, that this is the natural sleep pattern of humans.

The tradition of napping, an institution that the Romans called *hora sexta*, began its decline in 13th-century Europe due, it's believed, to the invention of the clock, although young children and elderly people still nap, and it's a very important aspect of many cultures.

Despite the stigmatisation of the nap as lazy or slothful in our busy modern-day lives, it is, in fact, a super-tool. The optimal duration is around 90 minutes, allowing us to go through all sleep cycles. But taking two naps of no longer than 30 minutes each – one in the morning and one in the afternoon – has also been shown to help decrease stress and offset the negative effects of sleep deprivation on the immune system. Even a 15–20-minute power nap has been shown to increase levels of alertness and boost mood, even in those who are getting a good amount of sleep at night. If you can't swing a half-hour nap during the workday, try grabbing a 20-minute siesta in your lunch hour. If you, like me, are a parent to small children, waking up every few hours at night, these short microsleeps can be helpful.

The timing of the nap is important as well. You don't want to nap too close to bedtime. Assuming you rise at dawn, aiming for between noon and 3 pm is sensible and consistent with cultures where napping is normal. And for those who may not be getting enough sleep at night or for athletes or those with a heavy life load, extending that nap to no more than two hours could be significant in making up some of that lost sleep at night.

Could napping ever be bad? Sleep experts are quick to point out that night-time sleep should be your priority. Only if you think naps are taking away from your evening sleep should they be avoided.

SLEEP HYGIENE FOR A STRESS-FREE SLEEP

Sticking to a routine isn't easy. But a daily awareness of some of the fundamentals of good sleep honours your natural biological rhythms and enables you to reflect on the day you just had: to consider what worked and what didn't, appreciate your accomplishments and shape a better tomorrow.

To summarise, combine the following with the de-stressing rituals in the next chapter to take back control of your day:

- Declutter your bedroom for a sense of calm, and plan a consistent bedtime in a cool and dark environment.
- Aim for seven to nine hours' night-time sleep; more if you are busy/stressed or chronically sleep deprived. Remember, quality is as important as quantity, so you might need to be physically in bed longer.
- Prioritise night-time sleep, but naps can be a useful tool if needed.
- Dedicate the final part of your evening to a screen-free wind-down (or block out the blue).
- Aim for a consistent wake time, followed by plenty of morning sun. Perhaps pause for a moment before diving on to social media.
- Don't hit snooze, as it disrupts quality sleep; just set your alarm for the time you actually have to get up.
- Avoid coffee after midday and be aware that while alcohol may seem like a fast track to passing out, it is actually a known sleep disturber.

WHEN SLEEP HYGIENE ISN'T WORKING ...

What if you have real trouble falling and/or staying asleep? Sleep hygiene is probably not going to be enough to shift the dial and increase the odds of you sleeping well. When you are actually in the throes of insomnia, sleep hygiene may become another source of worry and detract from the really big factors at play – sleep pressure and how aroused your mind is.

Cognitive behavioural therapy targeted at insomnia (CBTi) is an evidence-based treatment using techniques that address the mental factors associated with insomnia, such as the 'racing mind', and how to overcome the worry and other negative emotions that accompany the experience of being unable to sleep. In addition, CBT helps people with poor sleep establish a healthy sleep pattern. This behavioural element has been shown to be better than sleeping pills, supporting people to develop a 'pro-sleep' routine and to achieve a strong connection between bed and successful sleep, meaning that falling asleep and staying asleep in bed become more automatic and natural. If you struggle with sleep issues, it may be worth looking for someone who specialises in this area for support.

'SLEEPY' FOODS AND SUPPLEMENTS

Supplements might seem like an easy fix for sleep problems, but I'd recommend trying to incorporate some 'sleepy' foods first.

Melatonin meal

Melatonin is available as a supplement in some countries and on prescription, but there are ways in which we can improve our levels without pills.

Some foods contain tryptophan, the precursor to relaxing serotonin and sleep melatonin. Eating tryptophan-rich foods – such as dairy products, poultry and meat, as well as nuts, bananas, broccoli and spinach – can affect the production and levels of melatonin in the body. Tryptophan is also present in most protein-based foods or dietary proteins, including chocolate, oats and chickpeas. Be sure to combine with vitamin-B6-rich foods (avocados, beans, seeds, fish, meat) and magnesium from leafy greens to increase GABA (gamma-aminobutyric acid – a relaxing neurotransmitter).

Tart cherry juice contains melatonin itself and has a small beneficial effect on sleep patterns through improving its production. Pistachios are not just the most melatonin-rich nut, they are simply off the charts as the most melatonin-rich food recorded. For a physiological dose of melatonin, you only have to eat two!

Phosphatidylserine (PS)

PS is a chemical with widespread functions in the body. It can help with melatonin production by preventing hyperactive production of cortisol in the body, allowing unhealthy, elevated cortisol levels to decrease, and consequently more restful sleep to occur. We do get PS from our diets, mostly from animal products, but supplements are available and have been shown to be helpful with improving sleep. 100mg of phosphatidylserine, together with omega-3 fatty acid 3 times a day for 12 weeks has been demonstrated to regulate circadian rhythms and positively improve sleep quality.

L-theanine

This immune-strengthening compound found in black and green tea promotes relaxation and may aid sleep. Although these teas contain caffeine, L-theanine may interrupt the stimulating effects by helping your body process the caffeine differently. If, like me, you love coffee but hate that it gives you the jitters, add some L-theanine during your coffee kick or switch to tea for some gentler caffeination. It can also be a useful sleep supplement in the evening. A 2018 study found that people reported greater sleep satisfaction after taking 450–900mg of L-theanine daily for eight weeks. But always contact your healthcare provider first.

Glycine

Glycine is a well-known calming amino acid (a building block of protein) and helps the body make serotonin.[30] Taking 3g of glycine before bed helps with both falling asleep and quality of sleep by lowering your core body temperature and stimulating the release of gamma-aminobutyric acid (GABA), as shown in scientific studies.[31,32] It is the same neurotransmitter mimicked by some prescription sleep drugs like Ambien (zolpidem), and is also used by the body to make glutathione – a powerful antioxidant that protects and repairs our cells.[33]

Magnesium

Magnesium has a range of effects that calm the body and mind to prepare for sleep, working alongside serotonin and melatonin. You can get magnesium through drinking water and eating foods such as green vegetables, nuts, cereals, meat, fish and fruit. But even when eating a diet based on these nutrient-dense whole foods, it can be difficult to get enough magnesium. And we know that when we don't have enough, we can feel stressed and sleep poorly. Recent estimates suggest that both adults and kids in the UK do not reach recommended levels.

Very few studies have directly tested the effect of magnesium supplements on insomnia, making it hard to recommend specific amounts. But one way to relax your way to a restful sleep is with a warm magnesium salt bath, which has the added benefit of increasing blood flow to your extremities, cooling your core body temperature in preparation for sleep. Alternatively, taking magnesium in the form of magnesium glycinate (either as supplements or trans-dermal) has the added benefits of calming glycine.

TIRED BUT WIRED – IS CAFFEINE YOUR IMMUNITY'S FRIEND OR FOE?

Coffee's popularity can be attributed to a range of benefits that include keeping you awake and even triggering a feel-good 'dopamine' effect. Drinking coffee is also associated with lower levels of circulating inflammatory markers. Antioxidants in coffee are considered to help prevent heart disease, diabetes, osteoporosis and neurological diseases. Additionally, it is thought to cut cancer risk. But interpret these sensational headlines with caution – the research doesn't prove coffee reduces the risk of death. The link might just be down to coffee drinkers having healthier behaviours.

We do know that caffeine increases the stress hormone cortisol, which can not only impact sleep and cause insomnia, but in high doses can also cause anxiety, irritability and palpitations. Elevated cortisol levels are known to decrease the ability of your immune system to fight infections. Ethylamine – a metabolic by-product of L-theanine (see above) – primes the response of specific T cells that aid the body's defence against infection, allowing them to respond up to five times quicker. These immune-sharpening effects occur when caffeine is consumed in tea, not coffee, and in moderation – so balance is key.

But caffeine is the ultimate sleep prop. And caffeinism is a real thing. You wake to your alarm, feel a little low on energy, so you reach for a cup of coffee or tea to get you going. Morning caffeination is ingrained in our culture, but is this ritual depleting your body and zapping your energy instead of boosting it? Some people safely avoid this trap – they enjoy a morning cup and are good for the day. But the hectic pace of modern life demands that we switch ourselves on and off, and caffeine can service that 'On' switch when we need to stave off fatigue. So most of us aren't satisfied with just enjoying the one, and in attempting to maximise the good effects of caffeine, we push too far: exhausted from the day before and sleep-deprived, we hop on the caffeine cycle.

In general, more than 1.5g of caffeine a day can cause the typical symptoms of caffeinism (restlessness, nervousness, insomnia, increased urination and digestive disturbances). Generally speaking, up to 400mg of caffeine (roughly 3–4 cups/250ml of brewed coffee) is recognised as a safe or moderate amount for most healthy adults (adolescents should have no more than 100mg). But remember that the half-life of caffeine (time taken for the body to eliminate one half of the caffeine) varies widely between people, depending on factors such as age, body weight, pregnancy, certain medications and genetics. In healthy adults, the half-life is approximately five to six hours. So as a rule of thumb, aim for no caffeine after lunch. And if you, like my husband, argue that you can drink a double espresso after dinner and not feel any damaging effects on your sleep, you are mistaken – because even if you think you get enough sleep, its quality is most definitely impaired.

The effects of caffeine consumption in the morning when we have our peak CAR (cortisol awakening response levels – see p. 133) can actually diminish our natural morning cortisol and disturb the balance of our stress-axis. Worse still, by consuming caffeine at this time, your body is more likely to build a tolerance to the caffeine and the buzz you get will greatly decrease. And so begins the cycle of needing more. Cortisol does not only peak in the morning, it also fluctuates around lunchtime and early evening. Therefore, timing your 'coffee breaks' in between morning and lunch (e.g. between 9 and 11 am) takes advantage of the natural dip in your cortisol levels, so the traditional idea of a 'coffee break' makes a lot of sense.

HELLO SUNSHINE

Getting plenty of morning sunlight on your peepers is essential for keeping your circadian clock on good time, as well as for staying healthy through sunlight's many complex effects on the immune system. Vitamin D – the sunshine vitamin – has a long-established role in keeping our bones strong. But since the early 2000s, scientists have found hints that it is good for a lot more than just our skeletons. Vitamin D takes centre stage with nurturing strong immunity too. In fact, it controls the expression of over 1,000 genes. Yet vitamin D deficiency is rampant in modern life,[34] with around 50 per cent of people worldwide not meeting their daily requirement.[35]

About 10 per cent of the sun's output contains ultraviolet (UV) radiation which is necessary for vitamin D formation through a photosynthetic process in the skin. Scientists aren't sure exactly why vitamin D developed, but one theory is that it functioned as a kind of early sunscreen and, conveniently, it also helps our bodies to use calcium. It doesn't take much time to make enough vitamin D. The WHO suggest that 5–15 minutes of casual sun exposure without sunscreen on hands, face and uncovered arms a few times a week is more than sufficient to keep your vitamin D levels topped up. Fatty fish and fortified milk are a good dietary source too, if you eat those. But if you live on the wrong part of the globe at the wrong time of year, getting enough sun can be difficult. Factor in how old you are (a 70-year-old makes four times less vitamin D than someone in their 20s), how much body fat you have (vitamin D is fat-soluble, so having a larger amount of body fat means less is available for your body to use), your skin tone (pale people make vitamin D the fastest) and how easy it is to stay indoors all day, and you can see how getting enough sun without getting too much can be very challenging. This is a significant problem, particularly for your immunity.

Sunlight also Energises Your Immunity

Vitamin D has been shown to help the immune system fight off bacteria and viruses (another reason why we get sicker in winter). It's also emerging by association that plenty of vitamin D protects against a lengthy list of ailments, including multiple sclerosis, asthma, depression, heart disease and cancer. Research in Australia suggests that vitamin D deficiency is linked to an 11-fold increase in the likelihood of peanut allergy (and a tripling of the prevalence of egg allergy). Those with the lowest levels of the vitamin may have more than twice the risk of death from heart disease, compared with those with the highest levels.[36] Vitamin D has also been shown to be important for athletic performance and recovery.

In fact, when it comes to our immunity, vitamin D carries out a whole host of activities geared towards regulating all the various arms of this sophisticated system. For example, T cells, our master regulators, don't move around our body properly unless they detect enough vitamin D. Plus, it helps us produce antimicrobial substances in the skin, lungs and gut, which help defend the body against new infections. Additionally, vitamin D suppresses the immune responses by the inflammatory Th1 and Th17 cells, which are associated with many autoimmune conditions. Vitamin D also keeps our NK cells working well, which is a super-important factor in our bodies being vigilant against cancer and viruses.

According to one study, 77 per cent of patients who had an acute stroke were deficient in vitamin D. You can see why public-health messages actively promote vitamin D supplements to the public.

You want to sit higher up on the vitamin D curve, whether through supplements or getting plenty of sun-safe time outdoors. Since the vitamin lives in your fat stores for up to two months at a time, stockpiling sunshine during the late summer and early autumn could, theoretically, keep you healthy as the flu season heats up. Overexposure to the sun can lead to sunburn and even skin cancer, though, so with any amount of time you spend under the sun, sun block is key. But is it possible that all this sunscreen drastically reduces our skin's ability to make vitamin D? Despite some older studies suggesting that regular sunscreen wearers had only half as much vitamin D in their blood compared with those who didn't, more recent experiments indicate that wearing sunscreen makes very little difference to vitamin D levels. If anything, regular sunscreen users seem to have higher vitamin D levels – perhaps because, in practising safe sun, they get more time outdoors overall. This is a classic example of a confounder in scientific studies: good things happen in the sun – exercise, bike riding, walks with friends. People who get outdoors and, by association, keep their vitamin D levels topped up have healthier and happier outdoor habits.

In fact, there is a completely separate role of sunlight on immunity, not attributed to vitamin D. Specifically, the blue light found in sun rays makes disease-fighting T cells in the skin move faster, marking the first reported human cell responding to sunlight by speeding its pace.[37,38] T cells need to move to do their work. They need to reach the site of an infection from their lymph-node homes. Sunlight, as distinct from vitamin D, also helps with the development of Tregs – those oh-so-important regulators of unruly immunity. This is one of the reasons why many people find sunlight helps with skin conditions like eczema.[39] These are yet more examples of the immune system responding to the environment and a good reason why we, by design, are supposed to spend more time outdoors!

As the evidence continues to stack up, we need to realise that humans don't just need food, water and shelter, but sunlight too. For our bones, our immunity and, as you read in Chapter 3, for our microbiome.

IMMUNITY IS ROOTED IN THE SEASONS

Have you ever noticed that you crave different foods depending on the time of year, and that you also associate certain diseases with different seasons? The activity of the immune system changes from summer to winter and back to winter again. We don't just tend to suffer more runny noses in winter, but cold winter months make arthritis worse too. In fact, a whole host of health conditions are impacted by the changing seasons, including autoimmune diseases, allergies, heart attacks and strokes, and even depression, hip fractures, mental-health disorders, migraines and the effects of emergency surgery.

The winter increase in immune defences, designed to help stave off infections such as flu, unfortunately also raises the risk of harmful inflammation in the body, effectively lowering the threshold for heart attacks, stroke, diabetes and even some psychiatric conditions. Other conditions get worse during the hot months. In many people with multiple sclerosis, getting hot slows down the messages passing along nerves that have already been damaged by their disease, and this gives rise to increased symptoms.

Some infectious diseases are well known to follow a seasonal pattern, including human rhinovirus (the culprit behind the common cold) and influenza.[40] Another infection that has seasonal variance comes from *Staphylococcus aureus*, one of the more common causes of infection patients can get in hospitals, associated with food poisoning, skin disorders, pneumonia, meningitis and sinusitis. More infections occur in the summer and autumn, especially hospital-acquired MRSA.[41]

Not just our daily circadian rhythms but our seasonal immune cycles are a powerful part of our health. Evolution has selected for a seasonal variance in our immunity that is seen in changing gene expression at certain times of the year to reflect changing challenges. As much as 23 per cent of the human genome has some type of significant difference in expression in a seasonal pattern,[42] with genes switching on and off accordingly, depending on summer or winter, rainy or dry season and the types of infection that predominate in each.[43] Seasonal variation in gene expression, corresponding to temperature patterns and weather changes, impacts not only how well we fight infection but also the important regulatory functions that keep inflammation in check. Our comfortable modern world with the ability to heat homes all year round and enjoy bright light until late at night is confusing our immune systems and could now be a risk factor for inflammatory diseases that plague modern life. Perhaps the single-most important step to connecting with the seasons is to get out into nature. It doesn't have to take you far from home. Even if you live in the densest urban area, the natural world exists all around you.

A FINAL WORD ON SLEEP

Hopefully, by now it is clear that sleep, seasons and circadian rhythms are the foundations of our health, connecting our environment to our wellbeing via our immunity. No matter what your starting point to wellbeing is, I'd say sort out your sleep before anything else. Recognising the ways in which the changing day times and seasons affect you and looking at possible underlying reasons for this are simple, yet effective tools, empowering you to mitigate the impact of a busy 24/7 life.

CHAPTER 5

Mental Health Matters

'What we see in the microcosm is reflected in the larger organism: just as our cells need to stay connected to stay alive, we too need regular contact with friends, family and community. Personal relationships nourish our cells.'

DR SONDRA BARRETT, scientist and author, former faculty member at the University of California School of Medicine, San Francisco

Thoughts and emotions are an essential part of how we interact with our environment. Our brains and senses are constantly relaying information about what's going on around us and how we are feeling.

Have you ever felt butterflies in your stomach? A speeding heart rate in anticipation of giving a big presentation at work? These are physiological effects of your emotions. Not only do emotions drive your behaviours, but contrary to what we've long thought, psychology and immunology are intimately intertwined. I've already mentioned that immunity is your sixth sense (see p. 27). There is now a relatively new but growing field of psychoneuroimmunology – the study of how the mind can affect immune-system functioning – that is now starting to provide evidence for the interesting yet long-puzzling, ancient, anecdotally based notion that still sits on the fringes of mainstream science: that the mind can influence the body during health and disease.

The immune response is one of many 'tools' the body uses to adapt appropriately to life's ups and downs. Immune cells do not passively execute pre-wired defence mechanisms. Instead, they actively listen to outside signals – especially those coming from the brain – and act on them. It should be no surprise, then, that our nervous and immune systems speak a common biochemical language via shared molecules whizzing in both directions – messengers like hormones, neurotransmitters and cytokines enabling constant chatter between our immunity and psychology, each system with receptors not only for its own communication molecules but for each other's too. No system in the body ever works in silos. The cells in your body react to everything your mind says – negativity brings your immunity down. Every cell in your body is constantly listening to your thoughts.

The idea of emotions driving our health has long begged explanation, inviting untold scepticism from the medical community. Thanks to this 'immune–brain' axis, we learn to distinguish between what we like and what we do not like, to counteract a broad range of challenges and to adjust to our environment. Immunity must calculate the threat – what requires most attention? Infection fighting, tissue repair or running for your life? These intersecting systems help to establish the mind–body connection of health. Fail to channel your thoughts and emotions in the right direction and not only will your mood plummet but your immunity too.

However, I'd caution against overstating the power of mind over body in causing illness. This is one among a whole host of factors. There is no evidence, for example, that cancer is caused by negative moods or attitudes, and many illnesses may be unaffected by the neural changes set off by different emotions.

SICKNESS BEHAVIOURS AND THE IMMUNE–BRAIN AXIS

The immune and neuronal systems are the only two body systems that put a premium on adaptability. For example, anyone who has experienced an infection will recognise the psychological and behavioural components of feeling unwell. Just think about how you felt last time you had flu – your immune system dealing with the virus overlaps with depressive symptoms like social withdrawal, changes in appetite, inactivity, malaise, sleepiness, failure to concentrate and fatigue. These are clinically what we call 'sickness behaviours' – a suite of centrally organised behaviours that evolved for good reason.[1]

As the name suggests, sickness behaviours are designed to help us get over being sick, limiting us from doing more damage or spreading an infection any further. These adaptive changes are triggered by the immune system fighting infection, while the resulting immunological chemical messengers in the blood are received by the brain, instructing us to reorganise our priorities in favour of rest and recovery, to conserve our resources for the high energetic costs of fever and efficiently fighting infection.

Further evidence of the involvement of the immune system in brain health comes from recent findings that patients who survive sepsis – an overwhelming and life-threatening inflammatory immune response to infection – present later with long-term cognitive impairment, depression and anxiety as a result of the strong immune response they experienced.[2] Reciprocally, this neuro-immune crosstalk can flow both ways. Our central nervous system (CNS) can actively communicate with the immune system to control immune responses and vice versa. In other words, how we feel can affect how well our immune system responds to threats of danger like infection. And likewise, a raging immune response tells the brain to adapt our behaviour.

All this underscores the rising role of the immune system beyond infection fighting and disproves the idea that mental ill health is all in the mind. Of course, mental wellbeing does involve psychology, but it also involves equal parts of biology and physical health.

INFLAMMATION AND MENTAL HEALTH

Scientists have started to piece together that people with depression show classic sickness behaviour and sick people feel a lot like those with depression, exploring a common cause that may account for both. The immune system, or more specifically its inflammatory capabilities, is emerging as the *sine qua non* of mental ill health and a host of psychiatric conditions. Inflammatory cytokines and other features of inflammation have been shown to rocket during depressive episodes, and – in people with bipolar – to drop off in periods of remission. Healthy people can also be temporarily put into a depressed, anxious state when given a vaccine that causes a spike in inflammation.[3]

There are other clues, too: contrary to what was previously thought, people suffering from chronic inflammatory diseases such as rheumatoid arthritis know it's a long-term condition and can foresee that it will no doubt get inexorably worse until a very dismal end. However, their comorbid depressive symptoms are, in fact, a psychological outcome of sickness behaviours directly resulting from having raised inflammatory markers from their condition.

The link between immunity and psychology is a two-way street. We know, for example, that people who have had heart attacks as a result of inflammation in the arteries have a 50 per cent increase in depression.[4] And depression is a risk factor for heart disease and poor recovery from a heart attack.[5] Ultimately, inflammation is important to the body; it's our main defence from infection and not something we need to completely extinguish. But since inflammation by design is only ever meant to be a short-term assault, we don't have robust buffers in place to deal with it in the long term. There is also very much an individual 'sweet spot' or optimal level of inflammatory activity as regards to mood state.

Depression runs in families.[6] If your parents have been depressed, the chances that you have been or will be are significantly increased. But the answer to the question of nature versus nurture has remained obscure. A study published in 2018 was able to pinpoint for the first time a selection of genes associated with an increased risk of depression – a major milestone but one that also raises as many questions as it answers.[7] There were 44 genes in total, so quite a few, but each contributing a moderately small risk – so not a single binary genetic switch. With more research it's likely that this number of mental-health-related genes will rise further still.

So first, what are these genes? Many of them are, not surprisingly, important to the function of the nervous system; more surprisingly, many play a key role in the workings of the immune system, in particular its inflammatory response. And, second, why have these particular genetic variants that promote depression hung around through evolution? Well, until relatively recently (say, less than 100 years), infection was the major cause of death. These genetic variations that promote depression are thought to have arisen during evolution because they helped our ancestors fight infection.[8] Realistically, all of us will have inherited some of them, which places us on a continuous spectrum of risk, and our chances of becoming depressed will therefore depend in part on how many and the cumulative impact. This impact on the immune system is much more than the genes we are born with, but that makes sense. No other system in the body places such a premium on adaptability – except maybe the brain. If our immune systems don't respond to emotional adversity in our ever-changing environment, then we are dead ducks.

As the evidence continues to stack up, the obvious question is what might be causing the inflammation in the first place if you haven't got an overt chronic inflammatory disease like rheumatoid arthritis or heart disease or if you are not one of the unlucky ones with the perfect storm of mental-ill-health-susceptibility genes. What is causing the immune system to go awry and result in an inappropriate, depression-inducing inflammation?

Infection, as you have read, is not the only way to set off inflammation. Immunity isn't just hard-wired to detect germs, but also to sense danger. The immune system is not fully matured at birth. Our immunity needs certain inputs from our environment. The cells are there, but how they develop and how well they are regulated is very much nurtured by our encounters and adventures, shaped by our changing emotions and surroundings. Diet is no doubt important, but so are our microbiota, our ability to get adequate sleep and daily movement – ultimately influenced by our upbringing, environment and psychosocial status, plus the weight of each individual's invisible life load. It starts to look like mental ill health is a kind of autoimmune reaction to modern-day living, a normal response to an abnormal environment, going some way towards explaining it spiralling all over the world.

There have been no major advances in treatment for depression since the 1990s, despite it being the major single cause of medical disability in the world.[9,10] Recent history tells us if we want to make therapeutic breakthroughs in an area that remains incredibly important in terms of disability and suffering, we've got to think differently. Addressing mental-health functioning through the head only, using either talking therapy or antidepressants, for example (drugs that treat depression by manipulating neurotransmitters such as serotonin – selective serotonin reuptake inhibitors or SSRIs) has been, we are discovering, a very unhelpful way to look at it. But based on this new approach to the role of the immune system in mental health, there is now evolving work on anti-inflammatory agents as novel antidepressants.[11]

Anti-inflammatory agents may not have broad application, but rather help a subset of depressed patients who exhibit clinical signs of inflammation. We know that about one-third of depressed patients have consistently high levels of inflammation, and those with an overactive immune system are also less likely to respond to antidepressants.[12,13] The hope is that by measuring inflammation in the blood, drugs targeting the immune system will provide much-needed treatments for those patients who are non-responders to conventional treatments. There are already early clues in clinical trials that this approach is starting to work. Depression is a disease that affects hundreds of millions of people. Even if anti-inflammatory approaches help just a small proportion of them, that would still be a huge number. But if anything, the biggest impact may be on the way we think about the disease, making people less likely to believe sufferers should just 'pull themselves together'.

If we start to think differently, in the future an antidepressant won't just be a pill, but anything that lifts the weight of mental ill health. My hope is that more research in this area will mean that doctors listen to what's missing and gently take patients through the medical value of non-pharmacological interventions to restore those vital inputs. This could mean using food and lifestyle to keep inflammation in check, but also reinstating lost connections to people, our environment and our emotions.

But don't throw the antidepressants out just yet – there is now evidence that commonly used SSRIs that affect our brain chemistry may also affect the immune system in a good way. That's because immune cells need serotonin too. And by blocking serotonin reuptake, these drugs result in serotonin being used by key cells in the immune system to educate it to fight infection.[14,15]

AS ABOVE, SO BELOW: EMOTIONAL ROOTS OF DISEASE

Now you know of the link between immunity and depression, it shouldn't be surprising to hear that your state of mind can influence your immunity. Numerous scientific studies confirm the old adage that happiness and healthiness go hand in hand. Much of the medical community remains sceptical, but with piling evidence that virtually every ill – from the common cold to cancer and heart disease – is influenced, positively or negatively, by a person's emotional and mental state, we can no longer ignore this connection.

Hippocrates, the father of Western medicine, famously noted: 'It is far more important to know what person the disease has than what disease the person has.' Any doctor will tell you that two patients suffering from what seems to be the same disease will react very differently. Nearly a century after Sir William Osler (often referred to as the father of modern medicine) stated 'the care of tuberculosis depends more on what the patient has in his head, than what he has in his chest', scientists have started to decipher exactly how emotional states influence the onset and course of disease. Studies building evidence for emotional state in health are numerous and make a strong case for the immune system serving as both a channel and a controller of our emotional state, influenced by both our psychosocial and physical environment.

Like everything else in our lives, our emotions are mixed and impure, messy and, at times, contradictory. We experience anger, happiness, surprise and other emotions as clear and identifiable states of being: rising blood pressure in a moment of anger, gut-aching despair. Activation of brain regions associated with negative and positive emotions appears to weaken or strengthen people's immune response, respectively.[16]

Anger, often considered a negative emotion, is a fundamental part of our emotional state and not necessarily always bad. Triggered reflexively in response to a perceived threat or provocation (physical or emotional), it can direct behaviour in useful ways. If you're stuck in traffic, anger with the situation might motivate you to find an alternative route, which will then relieve your stress. However, it's less useful if you're in a similar situation, only stuck on a motorway with no option to divert. High levels of anger are associated with poor health,[17] changing the way immunity functions by exacerbating inflammation. Evolutionarily, anger preceded violence, so priming our immune system for action (inflammation) would be a short-term advantage. But when anger becomes persistent, it's problematic. And above-average levels of anger can even mean healing and recovering from illness significantly slower than others less disposed to this emotion.

Strong, positive emotional states have a distinctly beneficial effect on the immune system. Research suggests that those who engage in hobbies that nurture their sense of happiness and wellbeing (t'ai chi, art, breathwork) recover quicker from infection. Laughter therapy (simply put, watching a humorous film) has been reported to have a surprisingly potent modulating effect on the immune system, upregulating our virus-fighting and cancer-surveillance NK cells. Unfortunately, the attainment of happiness has helped to create an expectation that real life stubbornly refuses to deliver. But it's worth remembering that we are not meant to be consistently happy and perhaps this is something to be happy about. Pretending otherwise will only foster more feelings of anxiety.

LONELINESS – LIKE A BEE WITHOUT A HIVE

We are hard-wired to belong to a tribe, with an innate need for social connection. For a long time, a curious link between social isolation and health has been noticed. A number of illnesses, from heart disease to cancer to some neurodegenerative disorders, have been linked to loneliness. But with no studies directly testing this link, no one can really explain it. Some have suggested that through positive relationships you are more likely to engage in positive health behaviours, like regular exercise and eating well – known as the peer pressure theory of social relationships. But recent work in this area shows that bad habits aren't entirely to blame for worse health outcomes in lonely people.

Back in 2007, immunity emerged as the missing link between loneliness and ill health. The immune systems of lonely people seem to behave dramatically differently to those of the non-lonely.[18] Isolation fools the body into thinking it's in mortal danger, with increased activity of the sympathetic nervous system – responsible for the 'fight-or-flight' response – triggering inflammation without applying the built-in brakes that normally keep it in check. At the same time, when your brain sends a message that you are lonely, the immune cells hear 'Danger!', turning off some of your antiviral defences as your immune system diverts its resources, instead blindly pumping out inflammation. Unlike actual threats, over time these perceived threats start derailing immunity, via the slow-burn chronic inflammatory assault that raises the risk of a myriad diseases.

This is yet another clue that our social world interacts with our inner immunological one. Our immune cells have their own links to our psychology, which makes the world of the mind convert into the chemistry of the body. In the modern world, however, loneliness seldom equals mortal danger. You get lonely and feel rejected for all kinds of non-life-threatening reasons. But your immune system doesn't get the context in the 'danger' memo. No matter how trivial, it reads it as a danger signal to which it must respond. And this is important: we are constantly connected via the internet and social media to a degree never before imagined – it is the perceived rather than the objective social isolation that corrodes our health. Most estimates quote between 20 and 30 per cent of people in the UK as being chronically lonely. And among Londoners, that number is closer to 50 per cent. Social isolation is emerging as an established and robust risk factor for disease. Nothing can compare. So far, research suggests that the situations that most reliably predict a dangerous surge of inflammation involve social rejection, illness and loss: people suffering from trauma and cancer, those grieving or caring for dying spouses, and people dealing with relationship difficulties.

STRESS AND IMMUNITY: A SENSE OF DANGER

One aspect of our modern-day lives that plays into immunological disquiet is something we hear about and mindlessly mention in a negative light on a daily basis: stress. The term 'stress', as it is currently used, was coined in 1936 by Hans Selye, a pioneering Hungarian endocrinologist; he defined stress as 'the non-specific response of the body to any demand for change'. We only have one biological stress response, but an almost infinite number of causes (stressors). You can experience stress from your environment, your body and your thoughts. If we define stress as anything that alters our homeostasis, for good or for bad, then stress, in its many forms, is normal and vital for a healthy life.

So while stress mostly gets a bad rap, it's not all negative. 'Stress' is at best an ambiguous term. For some, it means excitement and challenge ('good stress'); for many others, it reflects an undesirable state of chronic fatigue, worry, frustration and inability to cope ('bad stress'). For the latter, I prefer the term 'stressed out', which conveys the chronic nature of a negative state. In essence, stress can be normal and appropriate when it is a short-lived 'acute' event – a gift from Mother Nature to help us, not kill us. Stress can be positive, keeping us alert, motivated and ready to avoid danger. Sure, you are not likely to hear someone saying, 'I'm really feeling stressed – isn't that great?' But if we didn't have some stress in our lives, we'd never grow as people, develop resilience and push ourselves.

The so-called 'good stress', what scientists refer to as 'eustress', is what we experience when we feel excited. Also called the 'fight-or-flight' response, it's a special branch of our body's autonomic nervous system – the unconscious control centre that has a built-in stress-response unit known as the sympathetic nervous system, initiating physiological changes to allow the body to combat stressful situations. Recent behavioural investigations suggest that acute stress can even encourage us to be more social and empathetic – useful for helping out the injured during an emergency, for example.[19] Through this we can experience wildly strong emotions and corresponding physical uproar in our bodies just by thinking thoughts. Most people will intuitively know butterflies in stomach, feeling their blood pumping, giddy but without threat or fear. It can make us feel alive and excited for life – like when we ride a roller coaster, get a promotion or go on a first date. Once the stressor has been dealt with, our parasympathetic nervous system (often referred to as the 'rest-and-digest') works in concert with the sympathetic branch to gently guide us back to baseline, our pre-stress healthy and happy set point.

THE SCIENCE BIT

Our central stress-response machinery includes the hypothalamic pituitary adrenal (HPA) axis and the sympathetic nervous system (SNS). Basically, it's an eloquent and dynamic intertwining of brain and body that uses hormones, neurotransmitters and immunity molecules. It's controlled by the teeny tiny almond-sized hypothalamus in the brain, which can't tell the difference between a little stress and full overwhelm. All it knows is that it's getting a clear neural signal from the amygdala – the brain's alarm system – communicating threats and strong emotions. Upon detection of danger, this involuntary (autonomic) network releases epinephrine (most often known as adrenaline – think adrenaline rush), which creates a kind of high, giddy, heart-pumping feeling, overcoming fatigue. The body is prepared for fight-or-flight within seconds.

In this time, immunity is also on red alert as we get a rush of inflammatory chemicals into the blood. Now inflammation is pretty important for defence, but it's also damaging and requires policing, so almost simultaneously there is the release of cortisol, a type of steroid hormone known as a glucocorticoid. This is the next critical step in the stress response, allowing us to survive moments of panic, as cortisol helps the body maintain essential functions, like blood flow, so we don't faint in pressurised situations. Collectively, this rush of stress also inhibits the action of the rational thinking brain, forcing us to fixate on the threat in hand.

Cortisol also curbs functions that are non-essential in the moment of panic. In fact, it dramatically affects immunity, with a wave-like dampening effect, reducing inflammation and the number of circulating immune cells. It also puts a pause on the generation of new immune cells. It might seem strange that stress should suppress our essential immune defences, but immune responses come with a substantial energetic cost. In that moment of imminent danger we need to focus our energy and all our reserves on surviving the threat to our lives. Infections, while dangerous, pale in comparison. Step one: survive certain death. Step two: deal with head cold. Hence, the result of this cortisol phase of the stress response is an inhibition of our immune defences and a prioritising of resolution, healing and repair.

WHEN THE NATURAL STRESS RESPONSE TURNS TOXIC

The stress response is a complex natural alarm system that is normal, healthy and, importantly, self-limiting. With a profound impact on our whole-body physiology, it is acute, short-term by design. Stress results from a specific event or situation – say, that split-second moment when you almost got into a car accident – and is the type we are designed to deal with. Once a perceived threat has passed, cortisol returns to normal and its impact on our body abates, leaving no lasting detriment to immunity.

This biological dynamic doesn't translate well into modern society, however. Modern life is stressful – not in the way our ancestors experienced stress, but partly because of the mismatch between the modern world and that which we evolved to encounter. The fight-or-flight response doesn't just come from external dangers, it also comes from within, via feelings of inadequacy or fear. Threats of death are relatively infrequent in real life, but our bodies are still triggered in the same way in the face of modern threats: running late for work, relationship stress, long working hours, social inequalities and so on. These threats to our wellbeing are real, but not quite on the same level as being chased by a hungry lion. And these modern-day stresses with no resolution can generate more stress. A stressful job or an unhappy home life can bring the heavy toll of chronic stress – what we normally think of as serious stress with the potential to impact our health. If your physiology is telling you to be constantly in fight-or-flight, then your tissues sense it as 'dis-ease'. You cannot heal and protect, when you are defending.

By evolution's standards, sustained 'chronic' stress is a fairly recent invention and our bodies aren't designed to cope with it. Consequently, we can face negative health effects (both physical and emotional) if it persists for an extended period. What is less clear, but highly relevant to modern-day life, is how the acute stress response can be activated by smaller sporadic and less obvious triggers, adding up to 'acute intermittent stress'. For example, screens and technology can activate our sympathetic nervous system, which is part of our biological stress response. Kids who use smartphones in the evening experience evidence of activated fight-or-flight and a reduced parasympathetic nervous system – the stress 'Off' switch. Sending an email and waiting on a response activates the sympathetic nervous system. How many of us are doing that all day?

When stressors are always present, and we constantly feel under attack, it not only erodes mental health but also immunity. Many of us get stuck in an unconscious stress that the body hasn't forgotten (even though we might have). Stress is perhaps the most overlooked and underappreciated factor in the likelihood of experiencing poor health, from a cold to chronic inflammatory conditions like heart disease, autoimmune disorders and even premature death.

THE NEW NORMAL

Stress becomes negative when we face continuous challenges without adequate recovery. Just as exercise is a stress that demands rest and relaxation between events for adequate performance, without downtime we become overworked, filled with stress-related chronic tension, causing wear and tear on the body – both physical and emotional. Every loop between social threat and immune response further sensitises the nervous system to danger signals. The whole stress response becomes derailed as the body becomes slightly deaf to the 'Stop' message and hypersensitive to the 'Go' message. Every burst of inflammation releases molecules that stimulate the same neuroimmune pipeline that produced it in the first place. This sets in motion yet another vicious cycle that continually reinforces inflammation in a process called 'biological embedding'. In this way, stress drives inflammation and inflammation starts to drive the stress response.

SUBJECTIVE STRESS – WHAT'S YOUR LIFE LOAD?

The subjective nature of stress makes it difficult to define and hard to measure. Studies of healthy undergraduate students found that those who reported high levels of psychological symptoms in response to stressful life events had only a third of the level of NK cells compared with students with the same number of life events but little psychological reaction to them. Studies of smokers who develop lung cancer compared with smokers who thus far are free of cancer have shown perceived stress to be the differentiator. However, you might argue that it is impossible to tell whether the emotions caused the disease or the disease caused the emotional reaction. Studies of people before they get ill are far more impressive. Observations in cancer patients with a high level of perceived stress found that immunological responses were significantly reduced even before the cancer developed.

While it might be hard to agree on what is and is not stressful, we can perhaps all agree that the sense of having little or no control in our lives is always distressful – and that's what stress is all about. When our response to stress becomes more damaging than the stress itself then it becomes a problem. And the more this happens, the more normal it can feel, not even registering as particularly stressful, just business as usual. In fact, only very recently a curveball to the concept of what stress is was proposed with a new theory: generalised unsafety theory of stress (GUTS), which takes into account perceived safety rather than actual threats. Generalised unsafety refers to many aspects of our modern lifestyles, including things like loneliness. The term 'allostatic load' has been coined to clarify ambiguities associated with the word 'stress'. I like to think about allostatic load as a cup – if your stress cup is full to overflowing for long periods of time, you are also generating the trickle-down immune-harming health effects of stress. It's important to keep in mind that we are all individuals with different issues filling the stress cup.

WHAT DOESN'T KILL YOU MAKES YOU SICKER

If you have ever been prescribed steroids by your doctor, you will be familiar with the strong immune-dampening effect and your doctor will likely have highlighted the potential long-term side effects of that 'Off' switch. Similarly, the longer or more frequently we are stressed, the more of this anti-inflammatory signal our immune cells will be receiving. When acute becomes chronic, cortisol ceases to be helpful. It forces our immune cells to be pulled back out of circulation and stored in immune tissues, instead of performing their important surveillance function. Over time, disruptions in immune-communication cytokines make immune cells less responsive to messages of infection. New immune cells are made at a lower level and ones that remain are less likely to function properly, accelerating the overall decline in our immunity, much like that seen in ageing (as discussed in Chapter 2). On top of that, our crucial cortisol-awakening response (see p. 132) cannot perform its vital function. This is a fairly significant derailing and one that can not only impact our ability to defend ourselves from infection and seek out potential cancers but can also make us more likely to respond inappropriately. Enter autoimmunity and allergy.

TOO MUCH STRESS LEAVES YOU OPEN TO INFECTION

During times of intense or frequent stress, immunity is no doubt disrupted across the board. However, it favours preserving antibacterial responses rather than antiviral ones. If you were in danger for your life, perhaps about to be hit by a bus, it's more useful to be able to fight off bacteria that may infect a potential wound than to prepare for antiviral responses against a cold or flu. This is reflected when this primitive system kicks in under modern-day stresses. More than 80 years ago, when Hans Selye made those first observations of the dramatic impact of cortisol on the immune system, lab rats were subjected to unpleasant stress and exhibited significant and profound immune suppression: their thymus glands decayed (the generator of T cells) and they got more frequent tumours, stomach ulcers and a host of different ailments in direct correlation to the stress they were subjected to. Since then, work in humans looking at the antiviral responses of students during times of exam stress found this a time when impaired antiviral immunity led to higher susceptibility to colds, coughs and flu. So getting ill might be the cost you pay for that stressful bit of last-minute cramming – turning on your sympathetic stress response, hijacking your immunity.

But it's not just about fighting off colds and flu. Viruses that belong to the herpes family can be a problem too. These include herpes simplex (which cause cold sores and genital warts), varicella-zoster (which causes chickenpox) and others including cytomegalovirus and Epstein-Barr virus. Most of us will have acquired one or other of these. For example, Epstein-Barr is thought to infect 95 per cent of the population. It is part of our total microbiome and mostly capable of living inside us for decades without causing problems. This is called latency. But these viruses are opportunistic and can switch to what is known as a 'lytic' stage, causing flare-ups. Yet they don't re-activate randomly; they wait until the immune system is distracted (e.g. by stress), taking the elevated cortisol in the body as a signal to enter the active lytic stage. While these viruses are able to take advantage of us being stressed, they are also a stealthy stressor on the body, stimulating the HPA axis to further elevate cortisol so they can continue to live inside us and overwhelm our immunity. Taking antiviral medications in and of itself won't provide a long-term solution if the stress isn't addressed.

STRESS AND ALLERGY

Although we don't think stress directly causes allergy, it seems to play a role, and for the most part it seems that stress can make allergy symptoms worse. Since the 1970s and 80s, numerous reports have shown that increased stress and anxiety go hand in hand with allergic flare-ups. In one study, 39 per cent of participants had more than one flare-up of allergy symptoms that correlated with periods of higher stress.[20] An allergic flare might not manifest immediately during a stressful incident, but may occur days after an increase to daily stress levels, meaning the link is often not obvious. Stressful situations can even increase the likelihood of an asthma hospital admission.[21] Family conflicts precede the development of asthma in kids[22,23] and psychological stress in pregnant women is increasingly considered a possible allergic programming agent.[24]

If you recall from Chapter 1, the highly specific adaptive immune response has several subdivisions. Cortisol cleverly shifts the balance to what's called the 'Th2', favouring an allergic response.[25] This explains why a portion of asthma patients can be resistant to glucocorticoids drugs. The good news is that while stress can make your allergy symptoms worse, relieving stress can strengthen your immune system and help you manage allergies better. One of the best ways is to not only identify your allergy triggers and seasons, but also your stress triggers.

STRESS AND AUTOIMMUNITY

Stress is often overlooked in the mosaic of autoimmunity, but their relationship is intricate. Stress is associated with both triggering disease onset and exacerbations of current conditions. Let's take the example of multiple sclerosis (MS), a devastating autoimmune disease of the nervous system. The incidence of MS has increased over the past 60 years, disproportionately so in women, from a 1:1 female:male ratio to more like 3:1.[26] This is, in part, attributed to the increased stress placed on women in modern-day life. In rats and chickens spontaneously developing autoimmunity, it has been found that they have underlying issues with their cortisol systems. Of all the immune-cell types, the T helper 17 cells are probably the most resistant to cortisol. Conveniently, they play a major role in driving inflamed autoimmune tissues.[27]

If you have been previously diagnosed with a stress-related psychiatric disorder, you have a higher risk of autoimmune diseases. And it seems the more severe the stress, the higher the likelihood. But for those worried that a stressful period at work could trigger an autoimmune disease, don't be. There is a big difference between being stressed and a diagnosed psychiatric disorder. The list of disorders that were tied to an increased risk for autoimmune diseases included post-traumatic stress disorder (PTSD) and serious traumatic experiences.[28,29,30,31] PTSD sufferers often have dysregulated cortisol, which plays into altered immunity and associated health problems down the line. Whether you have experienced PTSD or not, psychological therapy and cognitive behavioural therapy aimed to reduce stress levels have been shown to be effective in influencing better outcomes in many autoimmune diseases.

ADVERSE CHILDHOOD EVENTS (ACEs)

Studies now show that childhood stress can damage the immune system and have long-lasting effects on health. In the last couple of years there has been a lot more discussion around the role that ACEs, such as physical and emotional abuse, play in chronic illness. Scientists found that long after difficult events had passed, stress was still having a negative impact on children's immune systems well into adulthood, even if they had never felt depressed. Children react to stress slightly differently from adults. When a child encounters sudden or chronic adversity in their environment, stress hormones cause powerful changes in the body, altering its chemistry and hard-wiring how their DNA is read. Stressors are interpreted as fact, which leads to a hypersensitivity to stress subconsciously, even if their adult life is pretty good. The brain and body are still processing information from their environment as if it is a dangerous place, and stress response is disproportionate and perpetuating. The developing immune system and brain process information from the environment and react to this chemical barrage by permanently resetting children's stress response to 'High', which, in turn, can have a devastating impact on their mental and physical health as they grow up. What we see is increased chronic inflammation decades down the line, pushing their risk for autoimmune diseases. A child with four or more ACEs has double the risk for asthma as one with none. People with the same autoimmune disease, but no ACE had less risk for hospitalisation than those who had experienced ACEs; and hospitalisations increased with each ACE.

THE TRICKLE-DOWN EFFECT

Many people are unable to find a way to put the brakes on stress. Chronic low-level stress turbocharges our stress response, making that alarm bell ring a bit more readily. Stress today makes us a little more vulnerable to stress tomorrow. After a while, this has an effect on the body that contributes to the health problems associated with chronic stress. But it can also have a negative trickle-down impact, wreaking havoc on everything from diet to relationships.

When we feel unduly stressed, we are less likely to engage in healthy habits, we end up making bad decisions and have a harder time focusing. These can all impact on our health. In my own experience, stress prevented me from getting good-quality sleep. The science tells us this prevents the important stress 'Off' switch from working. When we sleep, the HPA axis and the sympathetic nervous system slow, reducing circulating cortisol, epinephrine and norepinephrine, and the parasympathetic rest-and-digest nervous system takes over. With poor sleep, this response can continue at daytime levels. Some of the immunity-impairing effects of poor sleep (see pp. 123–33), therefore, are in part due to the effect of continual activation of the stress response. It's why when we are stressed, we wake up feeling tired from poor-quality sleep and more stressed than before. All the time, immunity becomes further dampened as the cycle through more stress and less sleep spirals.

GUT FEELINGS

Your brain is not working alone to unravel your emotions; your gut too is intimately linked into the system via the vagus nerve. Your gut uses the vagus nerve (see p. 182) as a walkie-talkie to tell your brain how you're feeling, through electric impulses called 'action potentials'. Your gut feelings are very real. Disruptions to the HPA–stress axis not only impact immunity but also play havoc with the microbiota you met in Chapter 3. When people are feeling healthy, relaxed and safe, their gut microbiome communities generally work together harmoniously in a predictable symbiotic manner producing all the metabolites that nurture balance in the body. When they are under stress, gut microbiome communities become discombobulated and behave in ways that are unpredictable and vary from person to person.[32] When people are not experiencing psychological or physical distress, it is much easier to predict symbiotic and harmonious microbiome behaviour.

HAVE YOU GOT LEISURE SICKNESS ...

A Dutch research group coined the term 'leisure sickness' in 2001 to describe the supposed phenomenon in which certain people, particularly workers under pressure, fall ill as soon as they take a break. A number of explanations have been put forward to explain this so-called 'leisure sickness', each with a common red thread: stress!

The pre-holiday build-up with its associated increased workload and preparations, particularly for people in high-pressure jobs/lives, inflicts a psychological stress. We may be more susceptible to catching an infection in those stressful weeks leading up to a holiday, but if the immune system is also constantly being signalled by cortisol to suppress inflammation, we may not see any symptoms. Only when that stress-induced immune suppression lifts – the moment our holidays start – can the immune system provide the familiar symptoms of the lurgy, due to the combination of pathogens living in the body, along with the absence of the anti-inflammatory signal. In other words, you might fall ill before your holiday and only notice it when you start to relax. However, we can't exclude the possibility that change of routine, sleep patterns, jet lag, food and perhaps more alcohol than normal can also contribute. Even just sitting on an aeroplane a few seats away from someone with a cold can lead to an 80 per cent chance that you will catch it too.

... OR ARE YOU BURNT OUT?

The term 'burnout' was coined in the 1970s by American psychologist Herbert Freudenberger and is used to describe the consequences of severe stress in doctors and nurses, or those in care roles who, in helping others, would often end up 'burnt out' – exhausted, listless and unable to cope. Nowadays, burnout receives constant media attention. It even made the World Health Organization's International Class of Diseases as an Occupational Phenomenon in 2019.[33] No longer reserved for those in helping professions, or for the dark side of self-sacrificing your free time to care for loved ones, it can affect anyone, from stressed-out careerists and celebrities to overworked employees and homemakers. The term is applied to a set of symptoms, rather than a clearly defined illness. Since my own brush with burnout, I've been left wondering: if we are the most knowledgeable species on the planet, why do we get burnout?

SYMPTOMS OF BURNOUT

Strictly speaking there is no such diagnosis as burnout and medical professionals have yet to agree on what it actually is. But it is considered to have a wide range of symptoms:

- Feelings of depletion and exhaustion: sufferers feel drained and emotionally exhausted, unable to cope, tired and down and lacking energy. Physical symptoms include pain and stomach or bowel problems.
- Alienation from (work-related) activities: work becomes more and more stressful and frustrating. Sufferers may start being cynical about their working conditions and their colleagues. At the same time, they may increasingly distance themselves emotionally, and start feeling numb about their work.
- Reduced performance and professional efficacy: mainly affected are everyday tasks at work, at home or when caring for family members. People with burnout are very negative about their tasks, find it hard to concentrate, are listless and lack creativity.

Because the symptoms of burnout are similar to those of depression, some people may be wrongly diagnosed with burnout. So be very careful not to (self-) diagnose burnout too quickly. This could lead to unsuitable treatment. For instance, someone who is 'only' exhausted because of work can recover if they follow advice to take a longer holiday or time off work. But if someone with depression does so, it might actually make things worse because the kind of help they need is very different. Bottom line: if you are worried, talk to your health-care provider.

MIND–BODY MEDICINE

Our mental health affects how we think, feel and act. But it also impacts on our physical health, how our body deals with infection, how we recover and repair, and our susceptibility to and handling of disease.

Management of our mental wellbeing and treatment of mental-health disorders are at an important juncture. The current pill-popping prescription-focused model has achieved only modest benefits in addressing the burden of poor mental health worldwide. New breakthrough science linking the brain and the immune system bridges mind and body. Although the determinants of both physical and mental health are complex, I believe the future of mind–body medicine is bright. The role of the immune system as a crucial factor is compelling, and suggests that we look at physical and mental diseases in a different way. The links between our physical health and mental health can only be explored if we consider our 'whole' selves, from immune type to personality. Stressful events are a fact of life. I am not suggesting you walk away from your stressful life or resort to a suite of supplements to support your physiology. But you can take steps to manage the impact these events have on you. Just making a 1 per cent change to many things will add up. Lifestyle hacks – as outlined below – can help to interrupt the constant flow of stress chemistry and reduce inflammation. The reward for recognising your life load and learning to manage stress is peace of mind and perhaps a longer, healthier life.

HAPPINESS HACKS

Life will always have challenges and we all have vulnerabilities. Stress is a constant in today's fast-paced world and can jeopardise our health if left unchecked. And it's not going away any time soon. Resources like grit, gratitude and compassion are powerful tools for mind and body. And although we may not see it, these are the ones that we can have the most control over. There are also many other things that can be done to help us cope better with stress and dodge or improve a mental-health disorder. Conveniently, many of these overlap with avoiding triggering unruly inflammatory responses by the immune system – like a sensible lifestyle and remembering the knowns of poor health (not drinking too much, not smoking and not taking drugs). Getting good-quality sleep is also important, as is staying physically fit. (Importantly, mood-boosting exercise doesn't need to be intense, nor does it need to involve a gym. Just 10 minutes of moderate-intensity exercise is sufficient to improve mood[34] and there's very little evidence that going beyond 30 minutes provides any further gain in mood – so literally walking around the block each day!) But we also need to be mindful of our life load and how we may be missing vital emotional, social and environmental inputs crucial in hard-wiring us for health.

The good news is that the stress–immune reflex exists on a spectrum and works as part of a positive feedback loop. One end of this spectrum is characterised by GOOD STRESS or EUSTRESS, i.e. conditions of short-duration stress that may result in an immuno-enhancing state; the opposite end is characterised by BAD STRESS or DISTRESS, i.e. chronic or long-term stress that can result in dysregulation of immune function. The role of stress and its mechanisms have only relatively recently been studied, so we are still catching up with how to manage them. But there are ways we can manage our stress response to make us feel more like we are in the driving seat. It's my hope that you will incorporate the methods outlined below into your day-to-day life to find that balance between work and play. Each of these stress-busting suggestions is complementary and the sum of all of them is greater than their parts. All require regular practice for maximum benefits.

Ultimately, long-term management of stress is something that all us humans must continually address in an age of increased pressure. The key is to put back in what we take out. As always, I do not believe these ideas to be a magic bullet that will heal all ills and make you live for an extra hundred years. Of course, you should still pay attention to your doctor, and make sure any approach will not have any adverse effects on you or conflict with any existing conditions or medications you may be taking. However, they are another string to your bow. And I'm a big believer that by educating yourself and utilising modern science to sense-check the potential benefits, you can build up a matrix of simple, time- and cost-effective measures that, as a collective whole, can do absolute wonders for your health.

The Facts on Friends

Cultivating a strong social network of family and friends is very helpful for your health. Sounds obvious, but one of the most common complaints about modern life is loneliness. Studies have consistently shown that 1 in 10 of us is lonely, and a report by the Mental Health Foundation suggests that loneliness among young people is increasing. So how can we stop it?

Social change is a big part of it. Our working, living and community environments are changing, with opportunities to see family and make friends waning. We are 'virtually' more connected than ever, the internet having replaced the pub or cafe as a place of social exchange. But now we need to think more openly and creatively about how to deal with a cultural shift that could be harming our health.

There is no easy way to combat loneliness. But being aware and proactive is a good place to start. Taking the odd moment to look up from our screens, saying no to overcommitting, and being involved and participating in a 'non-virtual' network may be among some of the best evidence-backed ways to staying healthy. Maybe you have the potential to get involved in a community project? Could you attempt to strike up conversations with people at work you haven't talked to before. Be helpful when you see someone struggling in the street. Reach out to an old friend by calling instead of messaging? Try asking people about their personal passions, and mindfully listen while others speak. Limit your complaints – your tone can determine how people interact with you. This doesn't mean you can't talk about real and difficult issues, but the brain has an innate 'negativity bias' (it is simply built with a greater sensitivity to unpleasant news). There seems to be an ideal balance in healthy relationships that almost automatically regulates that between positive and negative.

Boost Your Vagabond

The vagus nerve (the longest nerve in the body, so called because it wanders, like a vagabond) is the longest nerve of our autonomic (unconscious) parasympathetic nervous system. The vagus nerve is essentially the queen of that system – aka the rest-and-digest or 'chill-out' part. The more we do things that 'stimulate' or activate it, like deep breathing, the more we banish the effects of the sympathetic nervous system – aka the fight-or-flight or 'do-something!' stress-releasing adrenaline/cortisol part.

It is responsible for controlling heart rate via electrical impulses to the heart, slowing our pulse. Every time we breathe in, the heart speeds the flow of oxygenated blood around the body; breathe out and heart rate slows. We may all have a vagus nerve, but not all vagus nerves are equal: some people have stronger vagal activity, which means their bodies can relax faster after being stressed. This variability in heart rate, known as HRV (or heart rate variability) is an indicator of vagal tone. A marker of physiological resilience and flexibility, it reflects the ability to adapt effectively to stress and environmental demands. To a certain extent, vagal tone is genetically predetermined – some of us are born luckier than others. But poor vagal tone can also be related to many lifestyle factors, such as traumatic experiences.

One of the vagus nerve's jobs is to reset the immune system, switching off inflammation after a stressful event. This nervous–immune connection is known as the inflammatory reflex – a direct communication link between the immune system's specialist cells in our organs and bloodstream and the electrical connections of the brain. Weak vagal tone means that not only is your stress regulation less effective, but inflammation can become excessive. Poor vagal tone is found in a range of chronic inflammatory diseases, with new research revealing that it may also be the missing link in treating chronic immune deregulation.[35] Vagus-nerve stimulation can significantly reduce inflammation and can be used to treat some cases of epilepsy and depression that don't respond to other treatments. Autoimmune conditions and many inflammatory disorders can be managed by vagus-nerve stimulation. Keep reading to find out how you can do this at home.

HRV represents a new way to track wellbeing. It's non-invasive and, although expensive measuring devices do exist, they are not necessary if you have a smartphone or chest-strap heart monitor (the ones people use in gyms or when running). The best thing is to check your HRV every morning and track for changes, looking for patterns as they associate with stresses and you incorporate healthier interventions. We know that people with a high vagal tone tend to be able to regulate inflammation and that this holds potential as a therapeutic strategy to blunt inflammatory disease. One study in which people were taught a meditation technique to promote feelings of goodwill towards themselves and others showed a significant rise in vagal tone. So doing things we enjoy and being around people who make us feel at ease are important. One of the simplest (and most sensory) ways to increase your vagal tone is through variations of breathwork and mind–body therapies, which I discuss next.

Mind–Body Therapy

Mind–body therapies (MBTs) – including t'ai chi, qigong, meditation, yoga and mindfulness – have been receiving more awareness from the mainstream, and scientific research is finally providing evidence that their benefits are real. As we struggle with our modern-day life load, MBTs are valuable tools for taking care of our wellbeing. They are also being incorporated into conventional medicine as a healthy lifestyle modification for people going through long-term treatment. My dad received reflexology while undergoing chemotherapy as part of his NHS care. Now there is no evidence to prove that reflexology – or any other complementary therapy – can cure or prevent any type of disease, including cancer. But there is some evidence that such therapies can help you relax and cope with stress, relieve pain and lift your mood, giving you that feeling of wellbeing when you need it most. They lower your heart rate and blood pressure, largely mediated by activating the vagus nerve. In those who are very unwell, it might just be the care and attention of having a therapist make them feel good. Touch is one of the most necessary sensory inputs that our bodies crave. The benefits of non-sexual touch have been scientifically proven to reduce stress, heart rate and blood pressure, and to help you feel connected through release of oxytocin.[36] Touch has even been found to lower the level of cortisol in the body (especially in women), protect from the common cold, and if you're already sick can improve symptoms and keep you from feeling worse. If you don't have another human to hug then similar effects have been shown with animals.[37]

Mindfulness meditation

This non-judgemental, in-the-moment awareness and acceptance of our thoughts and feelings has a calming effect on the mind as well as on our biology. This, in turn, diminishes the illusion that you have a threat to your life – like running from a wild animal – reducing inflammation and supporting your immunity.[38] Robust and persistent stress-management practices such as mindfulness are potentially the missing piece in moving the dial on your health with little risk. Meditation practice can be challenging. You might feel that it is just not for you or struggle with the cliché that it's another wellness fad – one better reserved for mystics and monks. Plus, when you are facing the demands of modern life, it's hard to find the time and discipline to do it daily, and difficult to know if you are 'doing it right'. Ironically, sometimes I find it quite stressful! But here are some tips:

- **Broaden the definition.** Sometimes we need to broaden the definition of meditation: consider the thought-numbing noise stopping you from being in the moment, from really focusing on how you feel and what you're doing – slow your mental chatter, stop frantically consulting your constant to-do list and take a quiet moment's pause to focus inwards.

- **Get comfy.** It doesn't need to be sitting for hours cross-legged in yoga pants. Choose a spot you enjoy being in and that's comfy. Sit up or lie down in a position that is relaxed, use support if you need it and wear something comfortable.
- **Get clear on your motivation.** If you know that meditation can help you cope with stress in your life, then start to view it through that lens as another tool in your toolbox.
- **Use tech to track your progress.** It's super-simple to start with popular apps like Headspace and Calm.com. And then, when it's time, there are tons of unexpected ways to go deeper.
- **Accept your thoughts.** It's normal for thoughts to come up, no matter how experienced you are. Practise calmly noticing and accepting them.
- **Write it down.** Research suggests that writing down thoughts and emotions can reduce day-to-day stress. If they're tough memories, it can lessen their hold; if they're positive, documenting them can increase feelings of gratitude. So break out that notebook!
- **Repeat, and stay accountable.** Remember, it's not about having a strict everyday practice but, after starting, the most important step is to repeat, repeat, repeat, making small changes consistent.

Small moments of introspection and mindfulness could be as simple as a non-podcast moment while commuting. It teaches us that it's attitude and emotions that determine experience and mood.

Breathwork

Breathing has long been observed to have an intimate link with mental functions. We can notice how our breath changes when we get excited. Conversely, voluntary slowing down or controlling the breath intermingles with many aspects of meditation practice. It is a particularly simple but effective practice for shutting off your sympathetic fight-or-flight response. And it's massively practical – a sort of meditation for those of us who can't meditate.

When it comes to effective vagal manoeuvres, any type of deep, slow diaphragmatic breathing – during which you visualise filling up the lower part of your lungs just above your belly button like a balloon and then exhaling slowly – is going to stimulate your vagus nerve. I'd suggest doing whatever type of diaphragmatic breathing fits your lifestyle and feels right. Working towards a longer out-breath than in-breath is what will activate your parasympathetic nervous system and improve your HRV. For example, start inhaling and exhaling to the count of six, and work towards a slower and longer out-breath.

While there are many different forms of deep-breathing exercises, box breathing can be a particularly helpful framework. Box breathing uses four simple steps, the aim being (as the name suggests) to help people visualise a box with four equal sides as they perform the exercise. This exercise can be implemented in a variety of circumstances and does not require a calm environment to be effective. Remember, practise regularly and track your HRV as an insight into your progress.

- **Step 1:** Breathe in through the nose for a count of 4.
- **Step 2:** Hold breath for a count of 4.
- **Step 3:** Breathe out for a count of 4.
- **Step 4:** Hold breath for a count of 4.
- **Repeat.**

Note that the length of the steps can be adjusted to accommodate the individual (e.g. two seconds instead of four for each step). Try it for three minutes to start. Consistent practice can have positive effects on multiple immune responses in a dose-dependent manner.[39]

Forest Bathing and Biophilia – The Gardener's High

I don't know about you, but I've always found something about being in nature – whether it's the sound of the ocean, the scent of a forest or an impressive countryside vista – can ease my stress and worry, help me to relax and think more clearly.

The idea that humans possess a deep biological need to connect with nature has been called 'biophilia', from the Greek, meaning 'love of life and the living world'. Turns out that despite millions flocking to live in built-up urban areas, we are hard-wired to affiliate with the natural world. The first hospitals in Europe were infirmaries in monastic communities where a garden was considered an essential part of the healing process.

Exposure to nature works primarily by lowering stress, rebalancing the sympathetic and parasympathetic arms. The Japanese art of '*shinrin-yoku*' (literally 'forest bathing') emerged in the early 1980s, recommending that people go out into Japan's extensive woodlands for the good of their health. Levels of cortisol drop and immune-cell function improves after walking in a forest for a few hours each day. Nature, it seems, is particularly strengthening for our NK cells (our main virus-fighting cells and cancer-surveillance system – see p. 11). Aromatic volatile substances called phytoncides from trees and plants – basically, that aroma you can smell when in the countryside or a forest – mediate these amazing health benefits. Perhaps there is an evolutionary need to draw us outdoors to top up our vitamin D levels too. People who spend more time in green spaces have significantly reduced risk for chronic diseases. Even a hospital window with a view improves healing, reduces the need for pain medication and aids the speed of recovery after surgery.[40,41]

In some countries, governments are translating this science into promoting nature experiences as a public-health policy. Therapeutic horticulture, community gardens and allotments are cropping up in urban spaces. 'Green exercise' in the presence of nature is reported to have better mood-boosting effects than exercising indoors. Only last year, NHS Scotland started a green prescription scheme for rambling and birdwatching. Biophilic design is now incorporated into workspace plans as the importance of the environment in health is increasingly understood.

Maybe that 'immune-boosting' magic we all desire lies in a little less time with our screens and more time staring at trees instead. If you can't get outdoors to decrease stress and assist with immune balance, why not diffuse woody phytoncide essential oils. Some of my favourites: Sacred Mountain, pine, cypress, Idaho balsam fir, palo santo.

Pavlovian Conditioning Rituals to Relax the Immune System

Perhaps you've heard of Pavlov's experiments with the dogs that began to salivate as soon as they saw or heard food-related cues because they learned they were about to get fed? Pavlovian 'conditioning' works because of the power of association. Now we might not be dogs – but the theory is still sound and can apply to many aspects of health, not least our modern-day immune systems.[42] In other words, our bodies respond whenever we see or think about something that has triggered a response in the past; and in the case of immunity, inflammation from a sense of danger can become a conditioned response, precipitating flares and making us feel unwell. The physiological mechanisms responsible for this 'learned immune response' are not yet fully understood, but in some way may tie into the placebo effect. Regardless, we can use this to our benefit.

Now we now know that immunity is tremendously affected by stress – and not in a good way. The solution is to break this learned association by creating new ones for relaxation – de-stressing rituals we can use as cues during times of stress to fight back against it and restore balance in our unbalanced lives. This is most helpful in a multi-step, multi-sensory way: a walk while listening to a relaxing piece of music; a daily bath or shower with a specific relaxing scent; a favourite herbal tea; and soft pyjamas before bed or diffusing a relaxing oil while listening to soothing music. Creating these associations and adapting them for morning or evening, combined with parasympathetic breath and meditation (see pp. 185–8) can help the body delineate the time of day in question – i.e. mentally preparing for the day ahead or unwinding in the evening. Importantly, it becomes a useful tool for interrupting the constant flow of sympathetic stress overdrive that is so pervasive in our daily lives. This has been used clinically in cases of allergy.[43]

Hormetic Habits

Learning how to manage stress is important, but just as valuable is learning how you can use stress to make yourself stronger.

There's a good kind of stress, called hormesis, that can make you more resilient and powerful in day-to-day life. It's when you push your body and it responds by becoming more resilient. While higher doses of that stressor would be toxic, exposure to mild levels stimulates the activation of the stress-resistance molecular machinery inside your cells. The stress that triggers these responses can also be known as eustress, as we learned earlier. This causes 'cross-adaptation', where one form of 'good' stress adapts the body for another, fuelling resilience against detrimental stresses in other areas of life.[44] Exercise is a classic example of hormesis – you damage muscle fibres and they build back stronger. But there are others too. Some of the most effective ways to stress your body and boost your resilience are through getting very hot or really cold.

What doesn't kill us might be good for immunity too. Remember, whenever you introduce a new health habit it takes time and consistency. But it only takes one bad day for you to find yourself right back where you began.

Cold Stress

Cold stresses the body in a good way. If you've ever popped outside on a cold winter's day without your coat, you'll know that your body reacts quickly: you start to shiver and your fingers, hands and toes turn blue. Cold-body therapy – known as cryotherapy – involves exposing the body to extremely cold temperatures for therapeutic purposes and is a type of hormetic stress. It has been used for hundreds of years in many different cultures (in Victorian times, cold baths were frequently prescribed for all manner of complaints from bruises to hysteria) but has particularly gained popularity in recent years. So, what is it about the cold that can make us feel well?

In the right doses, exposure to cold temperatures acts as a hormetic stressor that can encourage cross-adaptation to other forms of stress in your life. Immersion in cold water for those who are unaccustomed to it does not appear to yield benefits straight away. But with repeated immersions the body adjusts to the cold, sympathetic activity declines, while parasympathetic activity increases, improving vagal tone, which leads to a reduced, though not abolished, response to stress. We see evidence of cross-adaptation to cold stress in sports science where, compared with control participants, those who had been through a cold-adaptation programme were better at performing and recovering from exercise.[45,46]

There is a growing – albeit still rather small – body of evidence for cold benefiting our immunity. Cryotherapy can help reset our inflammatory response and reduce existing inflammation. Studies have shown that cryotherapy increases the presence of anti-inflammatory cytokines, as well as reducing pro-inflammatory cytokines.[47,48] In just 20 minutes of cold exposure, the stress neurotransmitters (adrenaline) can rise 200–300 per cent, inhibiting inflammatory cytokines, the known drivers of the inflammatory response, and reducing pain that accompanies many chronic immune-mediated diseases like arthritis.[49,50,51] Multiple studies have shown that cryotherapy reduced pain and improved joint mobility in rheumatoid-arthritis sufferers, with the effects lasting up to three months.[52,53,54] Consistent cold shock can increase immune cell numbers, in particular cytotoxic T cells responsible for killing cancer cells and virally infected cells. Cold-water immersion causes lymph vessels to contract, forcing the lymphatic system to pump lymph fluids throughout the body, flushing the waste out of the area. People who take regular cold showers were almost 30 per cent less likely to call in sick to work than others due to the improvements in immunity.[55]

The number of conditions associated with inflammation that may benefit from this intervention is huge. Steps are now being taken to investigate sea swimming as a therapy for chronic disease. Cold also enhances our internal antioxidants, which help us battle oxidative stress and excessive inflammation. Good for the brain, it encourages the production of feelgood beta-endorphins in the blood. The mind will focus 100 per cent on physical sensations and the brain feels like it has hit the 'Reset' button. Cryotherapy can help improve the quantity and quality of our sleep too. In turn, this helps the body to heal and recover, optimises cognitive function and helps keep our mental states balanced.[56] Importantly, it does not appear to have noticeable side effects or cause dependence, and it can be done for free!

Periods of cryotherapy can involve anything from a cold shower or a plunge pool, right through to high-tech cryochambers. Here's how to do it:

- **Start with a cold shower.** At the end of every shower, turn the water to the coldest setting and stand underneath for about 20 seconds. Build up to as long as you can stand. It's excellent for retuning your mindset after a long, stressful work day, as you transition from day to evening.
- **Upgrade to an ice bath.** This is the next level up from a cold shower. Use a plunge pool or prepare a bath with ice and see how long you can stand. Two minutes is a good target to work towards.
- **Whole Body Cryotherapy (WBC).** This involves short exposure to extreme cold via a cryochamber – a human-sized tank filled with liquid-nitrogen-cooled air. Exposure can vary from two to three minutes in temperatures that plummet to -130°C. Regular cryotherapy sessions can have a wide range of benefits, including reducing pain, reducing inflammation and speeding up recovery.[57]
- **If you live near water,** then just pop in for a weekly dip! Apart from the benefits of being out in nature, being a habitual sea swimmer decreases your likelihood of contracting a cold by 40 per cent.

Heat Shock

Stressing the body with passive heat – known as thermotherapy – as opposed to that generated by exercise has been used in various forms for thousands of years in many parts of the world for health purposes. Many cultures swear by the benefits of a hot bath, but sauna bathing is probably the most traditional. Used in Finland as a meeting point for pleasure and relaxation, saunas are now becoming increasingly popular in many other populations.

Frequent saunas can reduce the risk of a heart attack or stroke and can lower blood pressure.[58,59] Only recently has science begun to understand how passive heating from saunas improves health. Many of the benefits overlap with the stress-busting effects of the cold, but heat has some important and unique effects too.[60] Like cold, heat boosts the functioning of the stress-response system, which helps creates a virtuous cycle, helping the body to cope with stress and rebalance itself, via induction of heat-shock proteins. These molecules are made by all cells of the human body in response to stress to protect and repair our precious cellular machinery and are considered important for longevity. Heat-shock proteins are potent inducers of T cell regulation[61,62] and are themselves anti-inflammatory.[63] Through this, saunas can ameliorate inflammatory disease symptoms and benefit pain associated with chronic inflammation, such as that found in rheumatic disorders and other chronic-pain conditions.[64]

Regular sauna usage helps us prepare for and reduce oxidative stress by supporting our production of antioxidants such as glutathione.[65] In particular, regular sauna-goers have lower levels of chronic inflammatory indicators in the blood.[66] As well as fighting pain and inflammation, sauna usage strengthens immunity to infection, making us fitter, stronger and more resistant to getting sick in the first place, and more effective at fighting it off when we are. One of the areas where saunas have been shown to be important is in reducing the risk of respiratory infections such as coughs, colds, asthma and pneumonia.[67] When I had pneumonia, I visited the sauna almost every day. I found it the only place where I got relief. Sauna bathing has also been shown both to release feelgood endorphins and make us more sensitive to them.[68] Ultimately, this helps to put us in a better mood, and keep us there. If a sauna is not accessible, try regular hot baths. They may not be beneficial to the same extent, but can provide a feeling of well-being and relaxation. Many people like to stack a stint in the sauna with a plunge into cold water, amplifying the benefits of both. There is some evidence that this is scientifically synergistic.

Note: Anyone with underlying health concerns should always consult their healthcare practitioner before trying any type of hot or cold therapy.

CHAPTER 6

Intelligent Movement for Modern Life

'The principle is that the shape of a building or object [person] should be primarily based upon its intended function or purpose.'

LOUIS HENRY SULLIVAN, architect

Getting moving is one of the best things we can do for our bodies. It is the cornerstone of a proper lifestyle and is (thankfully) seen as indispensable for good health, not just for looking good. We can see the health benefits from population studies all over the world. But more than just weight loss or muscle gain, improving blood pressure or cholesterol, exercise has a profound, yet much less appreciated effect on our immunity, supporting it in crucial ways.

The benefits of exercise span from reducing infection and improving recovery to decreasing unwanted inflammation and strengthening your immunity.[1] But it doesn't stop there – exercise also induces valuable epigenetic changes that are inherited by your children.[2,3]

That being said, the relationship between exercise and immunity is a complicated one. It's a double-edged sword – dubbed the immunity–exercise paradox. There is a moment when a 'healthy' level of exercise shifts into 'too much' and can backfire on our health, such that we can actually improve our immunity by doing *less* of it. In the race of modern life, we need to take care of our immunity as an athlete would. So I've teamed up here with Emeritus Professor Michael Gleeson of the School of Sport, Exercise and Health Sciences at Loughborough University to provide you with the need-to know-answers to the good, the bad and the ugly concerning the relationship between exercise and immunity.

EXERCISE – NOT JUST ABOUT AESTHETICS?

For many people, exercise is mainly a tool to manage weight: fat loss and muscle gain are the desired outcomes of the weekly gym routine, plodding along the path to achieving a physique that conforms to our societal norms. But I'd argue that it should be more than that. Exercise is a critical component of the long game I introduced you to in Chapter 2. A healthy body mass is a foundation for strong immunity, which, as you have read, is the path to a long and healthy life.

Getting your body moving, *independent of weight loss*, can correct and even reverse some of the unhealthy changes accumulated through sedentary behaviour and provide symptomatic relief from certain chronic diseases.[4,5] Movement does wonders for your health, but one critical benefit is through your immune system. As I mentioned at the beginning of this book, some people seem to breeze through cold-and-flu season without so much as a sniffle. What's their secret? Well, if you look at the lifestyle factors that decrease the number of sick days, being physically active is perhaps among the most important. Moderate aerobic exercise – around 30 to 45 minutes a day of activities like walking, biking or running – can more than halve your risk of both catching a cold or flu and other common winter maladies in the first place, and of having a particularly severe form of them too.[6] Some cells, like NK cells, which deal with viral infections and cancer surveillance (see p. 11), increase tenfold after just one bout of activity.[7] Exercising a few minutes before a vaccination can even improve its protection.[8,9,10,11]

The consensus is that if you are someone who regularly participates in moderate aerobic exercise, you improve your chances of living better for longer. Regular exercise starts to influence immunity at the genetic level, deciding which genes are turned on to fight infection and control inflammation. Just walking on a treadmill at a moderate intensity has anti-inflammatory effects. This is one of the key ways in which being active lengthens health span, staving off not only chronic, low-grade inflammation, but age-related and chronic non-communicable disease. Put simply, physical activity is one of the best natural anti-inflammatory strategies we have – and with minor side effects, it just can't be replicated in a pill.

This makes sense from an evolutionary perspective – regular and robust physical activity is an intrinsic part of survival. We are designed to move; our immune cells have specific movement needs that are no longer satisfied by typical modern-day lifestyles. But while our physical bodies haven't changed that much over the past few thousand years, our lifestyles have, and we're moving too little and too infrequently. In fact, movement is almost optional in modern life. And not just movement, but posture too; the way we stand and sit alters our anatomy and physiology with far-reaching implications for our immunity.

If you need another reason to exercise regularly, try this: exercise can change the composition of your gut microbiome – in a good way. After six weeks of regular working out, the microbiotas of sedentary participants evolved to include more short-chain fatty-acid-producing bugs – the ones with specific immune-nourishing benefits. But after participants returned to six weeks of their normal sedentary lifestyle, their microbiotas reverted, so consistency is key.

In multiple discrete and intricate ways, underlying all these positives, exercise accomplishes all this by making our immunity stronger, keeping it in daily balance. If you become a lifelong mover, the benefits build up.

EXERCISE AND IMMUNITY

Whether you are sedentary, recreationally active or a high-level athlete, each and every single dose of physical activity has a profound and immediate impact on your immunity. And this instant response affects how well your white blood cells work, as well as the health of each individual immune cell.

Initially, during exercise, there is a mass exodus of white blood cells from our tissues into the bloodstream, as heart rate rises and blood pumps more forcefully. But then, within a few hours of exercise finishing, there is a sharp decline; these cells in the blood *appear* to disappear, typically falling to levels far lower than before the event for up to three days after a heavy session.[12] For decades, this convinced most researchers that a single hard long-distance race or other strenuous activity left the body so fatigued that it was unable to fight off cold viruses and other microbes. Heavy loads of physical exertion had killed large numbers of immune cells and created what some dubbed an 'open window' to infection – a few hours that allow opportunistic germs to creep in, unopposed. Marathon and ultra-marathon runners reported higher incidences of illness than the general population. But more recently, revised studies actually found that athletes are pretty poor at self-reporting illness, and, using more sophisticated measures, we now know that the risk following heavy exercise is no higher than the average. Fewer immune cells in the blood for several hours after exercise does not mean that these cells have been lost or destroyed. Instead, high-tech tracking revealed that they simply move elsewhere, migrating to the guts or lungs, areas that might be expected to need extra immune help after hard exercise (because, in the case of the lungs, faster and deeper breathing during exercise increases the chance of inhaling something infectious). A few immune cells also flow into the bone marrow, where they are thought to spark specialised stem cells into creating fresh new immune cells. So a small number of immune cells in the bloodstream in the hours after exercise is not immune suppression. Rather, immune cells, primed by exercise, are looking for infections in other parts of the body and moving to the tissues that are working hard, repairing our muscles.

At the same time, exercise also sends signals to our bodies to make new white blood cells – not just those infection-fighting immune cells, but also the Tregs (the master regulators, keeping the foot soldiers in line). Just as the rest of the body ages, so too do the immune cells. These old 'zombie' cells (we met them in Chapter 2) can be countered by exercise,[13] making it a fundamental tool for good health and balanced immunity. Out with the old and in with the new – over time, habitual exercise strengthens the whole immune system, keeping our white blood cells young and fresh.

MUSCLE IS YOUR IMMUNITY'S BEST FRIEND

Both muscle and fat are immune tissues in their own right. To fully understand the immune-strengthening benefits of exercise, we need to look at the intricate ways in which muscle and fat act immunologically. As well as being home to various immune cells, they produce their own versions of immunity molecules – adipokines (fat) and myokines (muscle).

As you read in Chapter 2, too much fat tissue in certain parts of the body, such as the belly, is a key contributor to low-grade inflammation – the kind that we don't want. If you are at an unhealthy weight, weight loss by exercise is way more effective in reducing low-grade unwanted inflammation than diet alone.[14] But as much as many loathe their body fat, too little can also be problematic since it plays a key role as a reservoir for important memory immune cells.

Immunity in the muscle is much more complicated. Despite inflammation's reputation as something inherently bad, it is required for your body to function and has many desirable effects. Athletes and coaches have long known that exercise can *cause* inflammation, and exercise-induced inflammation can be helpful. Training induces a considerable amount of inflammation in muscle tissue as a result of the damage to the muscle fibres. This inflammatory signalling initiates muscle repair and growth. Challenging our muscles through any form of physical activity causes formation of something called reactive oxygen species (ROS), often referred to as free radicals. Together with IL-6, these inflammatory signals collaborate, resulting in exercise-induced muscle damage (EIMD). Unlike problematic, low-grade, chronic inflammation, this muscular inflammatory response, driven by the release of large quantities of a particular immune-system communication molecule called interleukin 6 (IL-6) from muscle cells, occurs during and immediately after exercise, and is a valuable way for you to adapt to a physical challenge.

IL-6 can act as both a pro-inflammatory and an anti-inflammatory signal. Sounds confusing, but it depends on how much and where it's coming from – fat or muscle. High IL-6 in inactive people is bad news, an indicator of chronic inflammation. But IL-6 from exercise, though pro-inflammatory, is also a powerful switch, stimulating immune regulation. This not only helps clear out any unwanted inflammation, but also ensures that the threshold at which inflammation is triggered is kept at an appropriate level, limiting any accidental collateral from our inflammatory weaponry firing too easily.[15] So more intense and long workouts = more IL-6. And this has other benefits too: IL-6 molecules activate satellite cells – the precursor to muscles – strengthening and repairing our working muscles. IL-6 also plays an important role in the 'conversation' between fat and muscle, making it the reason why exercise is the best natural drug for treating metabolic disorders such as obesity and related complications such as type 2 diabetes.

Move Your Muscles – The Best Anti-Ageing Treatment

For all you readers over the age of 30, I've got some bad news: chances are you've already begun losing muscle. And it only gets worse. At mid-life, people start losing muscle mass and strength at a rate of 1–2 per cent per year, making it harder to carry out their normal activities. Up to half of those over 80 have thinner arms and legs than they did in their youth. In 1988 Tufts University's Irwin Rosenberg coined the term 'sarcopenia' from the Greek to describe this age-related lack (*penia*) of flesh (*sarx*) – muscle ageing.

Sarcopenia results from inactivity but can leave you more inclined to be less active, which itself leads to more muscle loss. This can occur at any age – a vicious cycle that will eventually lead to an increased risk of falls, loss of independence and even premature death. As we saw in Chapter 2, our immunity naturally wanes as we age. Thymic involution (the shrinking of our thymus gland, which produces our immune T cells) starts the fast track to weakened immunity in our 20s, while the slow burn of chronic inflammation increases with time. And up go risks for infections and damaging inflammatory conditions.

Enter physical activity, one of the most effective ways to tackle muscle loss and weak immunity in one go. Just as our immunity helps strengthen our muscles, moving them can rejuvenate our immunity. One recent research project examined 125 male and female cyclists, aged 55 to 79, who had maintained a high level of cycling throughout most of their adult lives. When the immune systems of older adults who had not done regular exercise and young 20-something sedentary people were compared with the active older cyclists', the latter's immunity knocked the other groups out the park.[16] Why?

Keeping your muscles active releases high levels of a hormone called interleukin 7 (IL-7) into the blood, which helps to stop the thymus shrinking, so it can keep producing fresh new T cells. Now the study above used older participants who had always been regular movers. But for the majority, the older we get, the less we move. According to one study, the most powerful 'deterrent' among the over-65s is a disbelief that exercise can enhance and/or lengthen life – the 'It's-too-late-for-me' excuse.[17] But research shows that adults who begin to work out in their later years still appear to live longer and have a lower risk of illness. You can train an older body and make a huge difference fast.

Movement not only improves immunity as we age, but also balance, strength, gait and muscle power, which helps prevent falls and injury. There are also big benefits for blood pressure and bone density. It's not just ageing that can bring about sarcopenia. A body weight higher than your healthy set point and/or a sedentary lifestyle can create double jeopardy, causing muscle wasting regardless of age. This is termed 'sarcopenic obesity' and has knock-on effects for overall metabolism too.

WHICH TYPE OF EXERCISE IS BEST AND HOW MUCH SHOULD WE DO?

Physical activity means different things to different people. Arguably, humans don't need to exercise, we just need movement. If someone does less than 30 minutes of physical activity a week they are defined as being 'physically inactive'. But physical inactivity isn't the same as being sedentary. Being sedentary for long periods of the day is bad for your health, even if you fulfil the criteria for being 'physically active'. So it's not about sitting at your desk all day then trying to counter it by jumping around in a gym class at the end of the day. But from cardio to anaerobic, strength and endurance, there are many different types of movement that can be performed to different intensities and durations. Which, if any, are best for immunity?

Broadly speaking, our bodies need two types of movement to perform at their best: aerobic and anaerobic. Aerobic means 'requiring oxygen'; generally light- to moderate-intensity activities like walking or swimming that can be performed for extended periods of time are aerobic. Anaerobic exercise is actually a misnomer because no form of exercise lasting more than a few seconds is 100 per cent anaerobic. But in general terms, anaerobic exercise is defined as 'lactate-inducing' from quick bursts of energy performed at maximum effort, like strength and resistance training or short-distance sprinting. Both aerobic and anaerobic exercise have significant positive effects on the cells and molecules of the immune system. Anaerobic activity has less impact on immune-cell function but does increase white-blood-cell count.

So how much should we do of each type? Government recommendations in the UK for those aged 18–65 are at least 150 minutes of moderate aerobic activity with at least two muscle-strengthening activities per week. But just walking one hour a week has been shown to improve health in the long term.[18] So if you are currently not moving much, this is a good place to start. Those who are more regular movers tend to be very loyal to one type of movement. But just as you wouldn't eat only one type of food, why limit yourself to one type of movement? Current data in exercise immunology suggests that regularly performing both anaerobic and aerobic exercise has the added benefit of improving immunity. High-intensity interval training (HIIT) ranks among the most popular current aerobic training methods, mainly due to its short duration, making it a time-effective workout for busy people. HIIT might have an impressive ability to increase your metabolic rate for hours after exercise, burning off more calories than traditional exercise for the same amount of time, but its beneficial impact on immunity is pretty unexplored. As well as being time-efficient, HIIT has a unique benefit over other less intense aerobic activity: it's a highly efficient way to encourage our immune cells to take care of their energy-producing batteries called mitochondria, rejuvenating them to keep them running well and fight off the ageing process.[19] Younger volunteers carrying out HIIT showed a 49 per cent increase in mitochondrial capacity and, even more impressively, the older group saw a 69 per cent increase. Lifting weights (or any kind of resistance work) is a hugely important part of exercise regimens, particularly as we age. Not only does it have significant functional benefits for healthy bones and joints, it's also important for our muscle mass, which steeply declines as we age. In fact, whether you are young or old, healthy or suffering from a chronic inflammatory disease like rheumatoid arthritis, muscle is your immunity's lifelong best friend (see p. 221).[20]

IS THE DAMAGE ALREADY DONE?

Exercise shapes up to be the magic miracle drug, and physical activity is still a largely overlooked yet modifiable risk-reducing factor, a countermeasure against persistent inflammation and the rising tide of inflammatory disease. So for a nation moving less than we used to, more worrying are the reasons why, despite guidelines and public-health messages, there is a growing 'knowing–doing' gap.

In the last century something unexpected happened: humans became sedentary. Sitting has become the new smoking, but while most people abandoned cigarettes long ago, how many of us have the luxury of ditching our desk chairs? One study revealed that most adults spend 15 hours a day sitting down, which, together with eight hours of sleep, leaves just one hour for physical activity. The modern environment is no longer naturally feeding into strong, flexible and aligned bodies. We need to pay special attention to how we move in the misaligned modern world in which we now live.

A common fear of many middle-aged folks is that they have missed the boat and they are too old to start reaping the benefits of exercise. And, yes, perhaps the best time to start exercising was 20 years ago – but the second-best time to start is now. Comparing a group in their 50s who had been training most of their lives to sedentary people who didn't start until they were at least 50 showed both can perform to the same level.[21] So everyone can benefit from exercise and it's never too late to start.

The relationship between sedentary behaviour and poor health is an ongoing topic of research that receives considerable media attention. Is sitting unhealthy for us primarily because we are not exercising when we are sitting? Or does sitting have its own unique negative effects on our bodies and, if so, could those outcomes somehow alter or even overpower the positive contributions of exercise? A recent small but worrying study suggests that sitting for most of the day could make us resistant to the usual benefits of exercise by altering our bodies in ways that are not just unhealthy on their own but also blunt the healthfulness of exercise. In other words, if we sit too much, our workouts may lose some of their expected punch. But that isn't a reason not to start; exercise has an undisputed role in our long term-health and wellbeing.

MOVEMENT RECOMMENDATIONS

As we've seen, the UK government currently recommends at least 150 minutes of moderate aerobic exercise plus two bouts of muscle strength training per week. (For reference, the modern hunter–gatherer tribe in Tanzania, the Hadza, do approximately 135 minutes of moderate-to-vigorous physical activity per day.) Only a third of the UK population is meeting government recommendations.[22] But even if we do fulfil the criteria for being active, we can still be sedentary, spending much of the day slumped in front of devices or sitting in cars.

The top barriers to engaging in any type of physical activity are time, energy and motivation – all in short supply as we work longer and harder than ever. But let's face it; if we don't move, the future isn't bright. As I said, it's less about smashing it out in the gym after a day in the office and more a case of just moving. All. Day. Long. Not exercising to achieve a specific goal but moving because we need to. How do we do that?

Physical activity can be manufactured all around you – from daily activities (like taking the stairs not the lift), to commuting (walking to the station rather than going on the bus), to using outdoor spaces (like the parks) or joining a parkrun. Your creativity is your only limitation. You could be extreme and join the latest wellness craze to 'rewild' your house and go 'furniture-free' (yes, it's a thing – the theory being that without chairs we are forced to squat and rise multiple times a day, the way our ancestors did, so we will never have back trouble, remaining limber into our old age). But if furniture-free is too daunting, just engineer your day so movement is a priority, keeping immunity strong all year round.

As a qualified fitness instructor, I love to get moving and generally enjoy exercise of all forms. But as a working mum of twins, I have little time for gym or classes. I find the struggle to keep active in the face of a busy schedule is frustrating. So I've flipped my mindset and searched out ways to move smarter. I try to vary my movement patterns like I vary my diet. And it's not as if we need a huge amont of variety to move and feel well – we simply need consistency and quality. Here are a few ideas (but don't feel you need to follow me or the latest trends; remember, this is simply about transitioning to a movement-rich life):

- **Future fitness.** Tech might have made our lives easier, but at a high price. It can nudge you to move, though. Stand-up desks, treadmill desks, Apple watches, Fitbit and apps to remind you when you have been sitting for a certain length of time are all helping to push more movement into day-to-day life.
- **Change your mindset.** Exercise doesn't need to be a burst of activity for one hour at the gym class. Think smart and look to move more throughout the day. Even if you take five to scroll social media, use that as a cue to sneak in more movement. Dreading doing all that housework? See it as meaningful movement. And best thing of all – it's free!
- **Consistency matters**. Plan, be organised and make space for movement throughout your day. You might not be

able to hit the gym at the same time each week but aim to be consistent in prioritising movement.

- **No gym, no problem.** Use your body weight for resistance and hop online to search for a quick and easy routine that you can do at home.
- **Listen to music.** Research shows that playing music while moving can help you improve performance, increase motivation and reduce distraction.
- **Walk and stand more.** According to research, being sedentary for most of the day – even if you do participate in exercise – increases your risk for poor health and immunity. Try a standing desk, walking meetings or cycling to work.
- **Start with your commute.** Park farther away, walk a stop or just take the stairs more often.
- **Walk and talk.** Meeting a friend or catching up with a colleague? Take your next brainstorm session for a spin around the block. Taking in the scenery while engaging in conversation can ramp up your inspiration.
- **Go green.** Get outdoors to move – whether it's for a walk, a class or playing with your kids in the park. Breathing in the air from outdoor environments provides a source of microbes to nourish your microbiota and, ergo, immunity.
- **Form follows function.** You might not live to exercise, but try to see movement as a way to maintain or improve quality of life and longevity. Functional movement prepares your body for daily tasks. Make sure to check you are moving with the correct form and posture in order to maximise the best functional end result.

LOVE YOUR LYMPH

One of the most important immune-strengthening benefits of movement is probably through the most neglected and least understood part of our body – the lymphatic system. This is the circulatory system of our immunity, a huge network of vessels and nodes that spans the entire body. Except for cartilage, nails and hair, our entire body is bathed in lymph fluid, called chyle. This clear fluid (of which we have around 15 litres, compared with approximately 5 litres of blood) permeates almost every nook and cranny of our bodies, carrying many of our immune cells, hormones and proteins – even mixing with the brain and spinal fluid.

The lymphatics have been historically the ugly stepchild of the body, somewhat neglected in favour of their fraternal twin – the blood circulatory system. While both systems share many functional, structural and anatomical similarities, the lymphatics are unique. Unlike the blood system, which is a closed loop with the heart actively pumping blood to oxygenate our tissues, lymphatics are open-ended. Movement through the network is instead governed by our rhythmic daily muscle movements, propelling chyle. In this way, physical activity contracts our muscles, forcing the lymph fluid through the body. For our immune system to function properly, our lymphatics – the grand avenues of our immunity – need to be in constant flow. So your foundation for health and wellbeing starts with the lymphatic system.

Lymph Is Your Lifeguard

If you don't own your lymph, you don't own your health. This magnificent network's purpose has long been misunderstood by the medical community, and its activities misinterpreted by most. In fact, if you look for information about the lymphatic system online, you'll mostly find references to 'swollen lymph nodes' and 'cancer', with relatively little in-depth information about the lymphatic system's function in healing and preventing illness. So let's start with a closer look at some of the core functions of lymphatics in our immune health.

Immune surveillance

As well as being the circulatory superhighways of the immune system, this transport around the body also fulfils one of the most important jobs of our immunity: surveillance.

Through the lymph, our white blood cells patrol all the remote corners of the body, keeping a lookout for infections, signs of cancer or anything untoward. The lymphatics act as immune-communication conduits, relaying information around the body, bringing immune cells together in hubs of activity called lymph nodes. You can actually feel some lymph nodes where they are close to the surface of your skin – on the sides of your neck, under your chin, under your arms and where your legs meet your torso. If you have ever had tonsillitis, then you're sure to have felt those unusually large lymph nodes, the tonsils. Surveillance of the body is a critical daily immune-cell task. When the lymphatic system is congested, and flow is sluggish, this has a negative knock-on effect on vital immune surveillance, and defence function can also be compromised.

Transporting dietary fats and fat-soluble vitamins

Your gut is full of lymphatic vessels that act as the entry point for fats and fat-soluble vitamins from your diet. From the digestive tract they transport them to every corner of the body; when the lymphatics are not flowing, you may feel energy levels drop and fat-soluble vitamins like A, D, E and K are poorly transported.

Detoxing

Like the drains in your house, the lymphatics are fundamental to filtering away waste products from the day-to-day running of the cells – toxic by-products from pesticides and environmental pollutants that are too big to enter the bloodstream. It picks up waste and, through an intricate series of processes, breaks them down and arranges for their elimination from the body via the liver. Waste products in and around our cells have the potential to age prematurely and lay the foundations for ill health. Your cells and organs can only be as healthy as the environment they swim in. So the lymph system is hard at work, 24 hours a day, in its relentless effort to keep the inside of the body cleansed and rejuvenated.

Maintaining whole-body fluid balance

We all know it's important to stay hydrated. As blood carries nutrients and oxygen around the body, fluid diffuses out into our tissues. One of the principal functions of the lymphatic system is to gather this fluid and return it to the blood system to maintain overall fluid balance. Swelling, known as lymphoedema, occurs when this fluid accumulates in a certain area of our body, such as a limb. Over time, persistent lymphoedema can lead to complications affecting the function of that body part, such as inflammation, fibrosis (a kind of scarring) and deposition of fatty tissue.

Note: Lymphoedema can be due to a specific genetic condition (known as primary lymphoedema) or due to surgery, certain infections and various lifestyle factors, including a lack of physical activity.

When Lymph Goes Wrong

Despite the many fundamental benefits of our lymphatic system, it can be exploited. Cancer cells can go into the small lymph vessels close to a tumour and travel into nearby lymph nodes. Here they can be destroyed, but some may survive and grow to form tumours in one or more lymph nodes. This lymphatic metastasis is a sneaky method employed by tumours to spread.

As we've seen, the lymphatics are the superhighways of the immune system, and because the latter is so vast and dynamic, it is especially vulnerable to change. Problems can occur when the daily flow and function of the lymphatics are disrupted. The damaging chemical storm of inflammation causes expansion of the lymphatic network and a weakening of its ability to do its job. This encourages the lymphatics to deposit fat tissue at the site of inflammation – for example, 'fat wrapping', that annoying tyre of tummy fat when you have an inflamed gut. These expanded leaky lymphatics don't drain so well, leaving us vulnerable to ill health and infection. Accumulation of fat tissue and inflammatory immune-cell infiltration are associated with the progression of the long-term inflammation that is implicated in several chronic health conditions.

The lymphatic system is uniquely susceptible to stress. Chronic exposure to large surges of cortisol, the stress hormone, can cause the lymphoid tissue to die off. Salt imbalance, poor digestion and an out-of-whack gut microbiome all impact the lymphatics, which intimately support our digestion and absorption of nutrients.

Lymph-Loving Lifestyle Tips

With all the ways in which the lymphatics can go wrong, being sedentary is probably the worst thing we can do for them. I can't think of any disease or condition of the body that can't be helped with effective lymphatic drainage. And our lymph is nourished by movement.

- **Get moving.** As my knowledge of the lymphatic system has grown, so has my interest in getting everyone to do as much lymph-loving movement as possible. But how much is enough? As a starting point, we know that sitting for too long increases our risk of dying, even if we exercise, so spend as much time as possible moving each day. As little as 20 to 30 minutes of aerobic exercise is enough to increase nitric oxide levels, giving a beneficial boost to lymphatic flow, particularly if you breathe through your nose. And the little things add up. Anyone who has ever worn a pedometer knows that you can rack up a healthy number of steps (studies show anywhere above 7,500 daily) just by moving throughout the day. And here is the kicker: slow or fast, young or old, overweight or otherwise – walking is what we are built to do and is enough!
- **Go swimming.** Water adds more beneficial pressure to those lymphatic vessels, so go swimming if you can (and even better if it's in cold water).
- **Try foam rolling.** This can be done with a foam-rolling device, which is a bit like a firm tube that you can use to perform self-massage. Foam rolling or any type of self-massage (by rolling a tennis ball over areas of tightness, for example) is great for increasing blood flow to the muscles, moving around lymphatic fluid and assisting the body to eliminate waste. It also helps to

break up tissue adhesions that can cause tightness and injury.

- **Practise deep breathing.** Just as the heart is the pump for the circulatory system, the diaphragm can assist in pumping the lymphatic system. Deep diaphragmatic breathing is the most important facilitator of lymphatic function. Combined with gentle stretching, this can also be a nice way to self-manage stress and relieve tension at the end of the day.
- **Bounce back.** Lymph fluid responds very well to G-forces, which is why mini-trampolines, often called 'rebounders', can be useful. Gentle up-and-down bouncing activates lymph flow. The gravitational pull caused by the bouncing causes the one-way lymphatic valves to open and close, moving the lymph and your immune cells all around the body.
- **Try dry skin brushing (effleurage).** Dry skin brushing naturally exfoliates to remove dead skin cells from pores, and clear the oil, dirt and residue that contribute to dull, dry, congested skin. The gentle pressure and movement of the bristles may also help stimulate lymph flow to gently detoxify the body. Dry skin brushing proponents also claim that it helps reduce cellulite by improving blood flow to the skin. To dry skin brush, you need a special brush with stiff bristles; begin at your feet and brush upwards with long, smooth strokes, always brushing the limbs towards the centre of your body. Do this along your legs and arms, then gently brush your stomach and back.
- **Go for a lymphatic massage.** While lymphatic-massage techniques may vary, it generally involves the practitioner manipulating the body to physically drain the lymphatic fluid, and it does produce tangible, evidence-based results. Often referred to as lymphatic

drainage, this was originally developed for the treatment of lymphoedema (see p. 213). A recent study showed that a combination of lymphatic-drainage massage and exercise was beneficial in the treatment of conditions involving blocked lymphatics following surgery. It is the only form of massage that has any scientific evidence pointing towards its effectiveness at increasing lymphatic flow and, therefore, improving the efficiency of our lymph system to do what it is designed to do.[23] Massage also reduces inflammation and helps aid recovery from injury. Plus, it's relaxing and can decrease stiffness and pain alongside a subjective release of stress.[24]

- **Alternate hot and cold therapy.** Lymphatic vessels contract when exposed to cold and dilate in response to heat. If you don't fancy running into the sea in winter or don't have access to a sauna, a hot and cold shower is a handy way to recreate lymphatic-nourishing responses at home.
- **Keep hydrated.** Lymph becomes thicker and less mobile when you are dehydrated, so make sure you drink whenever you're thirsty, and even when you're not.
- **Eat well.** Vegetables – notably leafy green vegetables and beets – contain nitrate, which converts in the body to nitric oxide. This, in turn, regulates lymphatic flow. Many plant foods, including fruits, the cacao found in chocolate (go for the dark one) and red wine, also provide polyphenols and other compounds that can increase nitric oxide production. High-protein foods such as nuts, beans, seeds, turkey, seafood and dairy products supply arginine, an amino acid used by cells to make nitric oxide.

HOW TO UP YOUR ROM

Range of motion, commonly abbreviated to ROM, is the full movement potential of our joints. Mobility is often associated with flexibility, but whereas flexibility is about a muscle's ability to lengthen in a passive state, mobility is about the body operating freely in its full range of motion – where muscles, tendons, joints and other components stay lubricated and move together actively and without pain. Most of us know mobility matters. If you struggle to touch your toes or get up from sitting on the floor or you experience restricted shoulder movement when reaching for objects, then chances are it's only going to get worse unless you do something about it. It's easy to take mobility for granted – until ageing, bad habits or injury cause you to have a lack of it, leading to back problems, missed work and sometimes surgery. Yet very few of us, even athletes, prioritise proper and full-range functional movement.

What may not be obvious, however, is that the effects of poor mobility go far beyond restricted movement. Problems with lymphatics can limit range of motion and vice versa, leaving us with muscle imbalances and tightness, susceptibility to injury and pain, and experiencing difficulty moving through day-to-day life, especially as we age.

Mobility is a direct determinant of things like posture and joint health and risk of injury. It influences how likely we are to participate not only in physical activity, but in life, particularly as we get older. In simple terms: use it or lose it.

There are few studies looking at what form of mobility conditioning is optimal in humans. Research on wild animals show that those with overall good body condition have the best immunity.[25] And you don't see many wild animals pumping iron in the gym or slumped at computers for hours. Using only your own body weight (calisthenics) is a great way to increase strength and fitness, with functional movements that will also improve your range of motion and flexibility. Plus, body-weight exercises tend to be inherently joint-friendly, not to mention safe and convenient. But mobility can also be just taking 5–10 minutes each day to pay attention to your range of motion and work on joint mobility with gentle stretches. Try a standing hamstring stretch, lunge with spinal twist or glute stretch. See how long you can squat each day and do a few cat-cow poses. Or roll down through your spine to touch your toes after waking each morning. Try performing the daily joint-mobility routine consisting of controlled articular rotations (CARs) as a ten-minute morning ritual. CARs are designed to help your joints stay healthy. During these movements, pressure expresses fluid and waste products, microcirculation brings in fresh lymph fluid and takes away the cellular waste. Everything feeds into your lymphatics.

RUNNER'S 'IMMUNITY' HIGH

You may have heard of the 'runner's high' experienced after exercise. This was originally thought to be caused by feelgood endorphins – the body's self-produced opiate drugs that relieve pain. But sometimes exercise also causes us some discomfort. This is due to dynorphins, also from the opioid family, that re-sensitise us to feelgood endorphins. It's not only these opioid effects that benefit the immune system, but also our exercise-induced endocannabinoids. These self-produced chemicals, similar to those found in marijuana, signal to the so-called 'reward centres' of the brain,[26] inducing a sense of sedate wellbeing – an altered state of consciousness that has long been appreciated by endurance athletes.[27]

The fact that we release reward-inducing, pain-relieving feelgood chemicals inside our bodies when exercising is thought to be an evolutionary motivation to stay active, helping promote the highly active lifestyle needed to dominate the environment. (When the going got tough, euphoria and pain reduction would have been invaluable.[28]) Does it sound like experiencing those feelgood chemicals could be a solution to the obesity crisis? Possibly not. Inactive people may not be fit enough to hit the exercise intensity that leads to this sort of rewarding sensation straight away, so taking those first steps on the path to regular activity will be challenging.

Feelgood chemicals are interwoven into the correct functioning of the immune system. They keep it running smoothly, acting like a feedback loop. These exercise-induced opioids help bring immune cells to the muscles being worked during exercise, helping to repair any damage. Without endorphins and endocannabinoids, the immune system begins misbehaving. This is one of the reasons why physical activity is important even when you are suffering from pain and chronic inflammatory conditions such as arthritis.[29] Studies indicate that it takes about an hour of endurance training for feelgood chemicals to be released. Overall, it's high-intensity activities, like short-term weight sessions or HIIT, that give the biggest high.[30,31]

A GOOD KIND OF STRESS?

Confronting our bodies with exercise is a form of stress. As we saw in Chapter 5, there are many paths to stress, but the biological reaction is the same: the fight-or-flight response via the sympathetic nervous system (SNS), co-ordinated by rises in neurotransmitters and hormones including epinephrine and cortisol. This reaction also acts as a signal that some form of stress is happening to the body, in the same way that psychological stress can, triggering us to adapt to the challenge, getting our heart rate pumping and mobilising energy to support our movement. In this way, exercise keeps our stress response 'well-oiled',' fine-tuned and ready for fight-or-flight.

In Chapter 5 you saw how the stress hormone cortisol acts like water to the inflammatory fire. Plus, moving our muscles also causes the release of important immune communication 'myokines' – IL-6, which help muscles strengthen, followed by anti-inflammatory interleukin 10 (IL-10). This means that regular exercise reduces unwanted inflammation, but also is a beneficial eustress, helping us 'cross-adapt' by politely asking the body to adopt better molecular coping strategies to new challenges and other stresses in our lives. These valuable adaptive processes are evolutionarily hard-wired into our genes. Myokines also stimulate our brain-resident immune cells (microglia) to produce growth factors that make new brain cells, protecting against neurodegenerative diseases. Movement can also help with serotonin production, which is important in our mental health too.

Exercise future-proofs us for new challenges and has, ultimately, allowed us, as a species, to cultivate resilience in the face of the unknown. When we continue to increase exercise intensity, our bodies respond by adapting, making incremental efforts to maintain the status quo.

A Word of Caution

There is no doubt that exercise sharpens the stress-response switch. But it must be exquisitely calibrated. Where the balance of power falls can make all the difference. Too much from a heavy training schedule or prolonged strenuous exercise like running a marathon has the biggest impact on exaggerating this natural anti-inflammatory response to exercise. Studies in both human and animal models demonstrate that too much exercise means the stress response is amplified far beyond usual levels for too long,[32] suppressing our immunity, disarming our vital defences. This drastically reduces circulation of the immune cells and antibodies the body desperately needs to fight off foreign invaders. Effectively, it tips the scales in favour of infection. In fact, it can mean higher rates of falling ill than in people who are inactive. This is seen in the J-Shaped Curve, a model first described in the early 1990s to express the relationship between exercise and infection risk. Moderate regular movers reduce their risk of infection compared to their sedentary counterparts. Bizarrely, the risk of both catching an illness and of becoming especially sick increases if you exercise intensely or for a prolonged period of time.

Prolonged release of cortisol particularly stymies NK cells – those that fight viruses and help keep tumour cells from gaining a foothold. Athletes who play hard often come down with more severe infections, and doubling your usual training load for a week can actually depress immunity by up to 50 per cent.[33,34] We know that frequent moderate activity is associated with reduced risk for several cancers, but endurance training – maintained for many years – may increase it for select cancer forms.

Training of any type can stress our systems. What if your relentless exercise is triggering greater muscle damage than your body can cope with? Exercise adaptation costs real energy and physical resources, of which there isn't an endless supply. When food intake does not meet the demands placed on the body, it has to carefully triage, borrowing resources from somewhere – and that somewhere includes the immune system, leaving you tired and pretty unhealthy.

If you have been doing too much strenuous or prolonged exercise or just have a dangerously large appetite for doing HIIT workouts (which can cause greater muscle damage than long sessions) on top of a busy schedule, this can increase your likelihood of developing an illness. Especially if you do it day after day, without recovery days – then the effects can be cumulative and immune function becomes progressively weaker. This can force you to step away from training routines that you love for longer than you would like, and everything you have worked to achieve will be out the window. But it's not just your performance that's affected. Your health can be seriously compromised without adequate rest and recovery. There is evidence from athletes who continued to exercise while suffering from flu that this can lead to a form of chronic fatigue syndrome, where the virus spreads throughout the body in a subclinical form, constantly engaging the immune system and causing exhaustion for a prolonged time.[35] Most of the studies mentioned on the negative effects of too much exercise on immunity have to do with aerobic activity. But since pretty much all physical activity will involve some form of aerobic activity, this is not just a result of chronic cardio.

Routinely overtraining before you have recovered from your previous workout causes a state known in exercise terms as 'overreaching' – fatiguing your muscles, your metabolism and your stress axis. As a tactic used strategically by coaches with athletes to improve their performance, overreaching isn't all bad and can be 'functional', in that it's a temporary state which, with adequate recovery, can improve future performance. However, intense training over an extended period without sufficient balance between training and recovery leads to 'non-functional overreaching', an overtraining syndrome that can last for weeks or months with significant consequences for immunity.[36]

What is clear is that functional overreaching, whether cardio or otherwise, can easily tip over to non-functional – and then exercise really isn't doing you any favours. The greatest harm to immunity is by training to exhaustion without giving your body time to recover. Keeping yourself well enough to show up to train is an important part of the whole long-term health equation. And these damaging effects are not just limited to our immunity and gym performance. They can also cause gut damage, dubbed 'exercise-induced gastrointestinal syndrome'. The risk of gut 'leakiness' and impaired digestive function, which can compromise your health and wellbeing, also increases along with the intensity and duration of exercise.[37] Exercise-induced pain, exhaustion or injury can be an added psychological stress on top of the modern-day life load. And remember that relative energy deficiency in sport (RED-S) is a serious illness with potentially lifelong consequences, arising from training with low energy availability (through consuming too few calories). This can result in fatigue, hair loss, cold hands and feet, dry skin, noticeable weight loss, increased healing time from injuries (e.g., lingering bruises), increased incidence of bone fracture, depression and higher susceptibility to infections.

SO IS EXERCISE HELPING OR HURTING YOUR HEALTH?

It's normally pretty simple: don't overtrain and don't ramp it up too quickly (long, high-intensity workouts with inadequate recovery). Do the opposite: shorter workouts, less often and make sure you're recovering sufficiently in between. But exercise-induced immune fatigue exists on a continuum. And it may creep up on you before you realise it. Are there any tell-tale signs that your exercise routine is harming rather than helping your health?

Generally speaking, the more strenuous the exercise, the longer it takes for the immune system to return to normal. The strain on the body is dependent on how hard, how long and how often. Exceeding what your body can deal with is the main culprit for immune depression and all the associated health risks – generally, long workouts of over one and a half hours, repeated bouts of intense exercise on the same day or over several consecutive days or training intensely for one to two weeks.[38] The tipping point is different for everyone, but your body will let you know when you've gone too far – you just have to listen. Some people can just 'feel it' when they could use a rest; others use various metrics (their lifting log, their mileage count, their coach's suggestions) or they simply take a rest day every so often, regardless of whether they 'feel it' or not. Here are some tell-tale signs of overreaching:

- You're feeling physically or emotionally exhausted from your workouts or repeatedly fail to complete your normal workout.
- You normally enjoy workouts, but start dreading the gym, plying yourself with coffee every morning to stagger on to the treadmill.
- You are losing leanness, despite increased exercise. Working out too much can cause muscle wasting and encourage fat deposition – you're 'burning calories' probably more than ever before, but doing too much limits resources, so the body turns to precious muscle tissue. Plus, too much cortisol causes increased insulin resistance.
- You are hitting it hard every single day, but with a hefty life load. Trying to maintain an intense physical schedule means recovery is compromised (especially if you have many commitments, a family and/or a job).
- You feel overly fatigued, sluggish and useless, constantly nagged by little aches and pains that never go away and feeling crap after every workout. Exercise generally elevates mood, so if it's having a negative effect on your mood, it's probably too much.

- You are falling ill more often. If diet is on point and sleep has never been better but you're still getting sick, your immune system is suffering from the added stress of overtraining.

Athletes train hard – almost every day, often twice a day – in order to be elite. Does this mean they are constantly in a state of DOMS (delayed-onset muscle soreness – those tender and aching muscles one or two days post-workout)? Fighting off non-functional overreaching to achieve their training goals? It's the progressive and appropriate amount of exercise overloading that helps to achieve the specific adaptations required for optimal performance, with planned-for rest and recovery, which is when adaptations occur. The key is consistency. Those first-timers filling up the gyms in January, but who don't make it a regular practice will always be trying to calibrate the most effective way to adapt to and recover from exercise-induced muscle inflammation. But it can take around four to six weeks' consistency for this to happen, as the body tries to determine how best to produce the energy needed both to support the immune system and fuel exercise and recovery. So take your time, and tackle exercise with a gradual progression of intensity to allow your body to adapt and your immune system to strengthen.

From putting immune function under the microscope like this, we can clearly see the J-shaped benefits' curve of exercise: some is good, more is not always better and too much is very bad. For dedicated gym-goers who are used to the no-excuses, just-do-it model of training every day whether they feel like it or not, this can be hard to hear. But the evidence is clear: this simply isn't good for you. It's not good for your immunity; and, over time, it's not good for performance either.

RECOVER HARDER THAN YOU TRAIN

If ever there was a magic pill, then prioritising recovery over training would be it. But sometimes you have (or want) to train hard. So what's the answer?

It's not simply a big total training load that depletes immunity, leaving it susceptible to attack, but rather how abruptly your training ramps up and how little recovery time you allow.[39] In fact, experts have uncovered that dramatic increases in training volume are perhaps a better predictor of catching an infection than just your training load alone.[40] Rest is arguably the most important factor in your recovery. Sleep quality and quantity are essential mediators. Many professional athletes sleep 10–12 hours a night to allow their systems adequate time to repair tissue, restore mental acuity, improve reaction times and optimise energy utilisation. You may not be an athlete, but getting more sleep can help if you are working out like one while holding down a full-time job and a heavy life load.

Restorative practices, such as manual soft-tissue work by a physical therapist or massage therapist, as well as compression technology,[41] play an important role in bolstering recovery and restoring immunity. Both cryo- and thermotherapy (see pp. 191–5) can also prevent muscle soreness and put you on the road to recovery.[42,43] In fact, regular sauna use improves your ability to work hard in the gym before you hit that point of no return.

There is some evidence that diet shows promise in countering the immune-weakening effects of overtraining.[44] With nutrition, it is very important to avoid any deficiencies, particularly protein and essential micronutrients (see Chapter 7 for more details). The jury is out on fat. But don't under-eat carbs. Training low (low-carb – either in a fasted state six hours or more after your last meal or twice in one day or on a very low-carb diet) is a strategy known to enhance training 'stress'. This might, at first, improve performance, but over time can erode it, along with your health. Making sure you maintain your carbohydrate levels as much as possible during and after exercise is the only evidence-based way to take the edge off exercise-induced fatigue and prolonged recovery. In fact, despite the popularity of low-carb diets, low carb is not recommended when you want to perform your best. For most of us, that means taking particular care to have carbs during high-intensity or prolonged exercise that lasts for 90 minutes or more. Between 30g and 60g of carbohydrates in the first few hours immediately after strenuous exercise helps to restore immune function.

Next up among helpful nutrients for your recovery is vitamin D. Overworking muscles can drive down vitamin D levels, leading to a deficiency that can impair muscle function and recovery. Plus, if you're deficient in vitamin D, you are three to four times more likely to catch a cold.

With overtraining, you need to realise that this is only part of the picture. Other factors like minimising psychological stress can be just as important. Exercise stress, like psychological stress, can also bring anxiety and disturbed sleep. Cortisol and inflammation can work against your gains. Overtraining can tax both the sympathetic fight-or-flight and parasympathetic rest-and-digest arms of the nervous system.

KEY TAKEAWAYS

- Our immune system is directly affected by each individual bout of exercise.
- The benefits add up: long-term moderate exercisers have the strongest immunity.
- High-intensity and/or for longer durations (1.5+ hours) without adequate rest can cause lasting immune depression and compromise your health.
- Rest, recovery and general lifestyle help prevent overreaching: carbs, sleep, restorative practices (massage, flexibility, mobility, restorative exercises) are all important.

PRESCRIPTION: BE EXERCISE SAVVY, WHATEVER YOUR LEVEL

I don't know you. I don't know your fitness level, how you slept last night, how much time you have, how full of stress your life is. But *you* know you. And you're now equipped with the knowledge of how to maximise exercise for your health without overdoing it: regular and moderate, aerobic and anaerobic is the secret for strongest immunity, reducing disease risk and optimising well-rounded fitness.

Ultimately, what matters most is moving, so the best workout will be relative – it's the one that gets you to show up every day (or the most amount of times possible per week). The aim is to be a recreational athlete for life:

- **LEVEL 1: Just get moving.** No energy, time-poor and feel like you have a long road ahead of you? Just start. And start small, keeping it fun and playing around with different forms of movement.
- **LEVEL 2: Challenge yourself.** The type, intensity, duration and frequency of exercise and the conditions in which it will nourish immunity, encourage adaptation and effectively reduce life stress are likely to be different strokes for different folks. So running could be a 'de-stressor' for some, while others would benefit from aerobics, swimming, dancing or yoga. But if your workouts are no longer challenging you like they used to (because of inflammatory adaptation), get over that plateau by changing it up. Varying your workouts between aerobic (low- and high-intensity), strength and mobility in your weekly routine is a great place to start.
- **LEVEL 3: Don't overdo it and get professional guidance.** If you're blessed with a lot of discipline and willpower, already spinning, doing yoga or CrossFit or running several times a week, it might benefit you to add a little variation into your weekly workout routine. Work on that push-up or pull-up until you can execute it with perfect form. And remember, you are doing your immune system a favour by not over-exercising.

MICRO-WORKOUTS – BANKING BENEFITS WITH LESS STRESS

One emerging fitness strategy that I love is that of micro-workouts – short 10-minute workouts comprising a mixture of strength-based and cardio, rather than a single official one per day. These deliver distinct and awesome benefits; several micro-workouts throughout the day add up to an incredible cumulative training effect without disturbing the stress/rest balance. And if you have a sedentary job, it's definitely an upgrade from the health hazard that is sitting down for long periods of time.

I'M SICK – SHOULD I EXERCISE?

Immune responses are costly, demanding energy and a triage of resources to fuel the infection-fighting inflammatory fire. If you don't have severe symptoms or are over the worst of it, but generally feeling depleted, there is little benefit in a heart-pumping, calorie-burning workout. On the other hand, a mild workout like a walk in the fresh air can help speed up recovery, propping up the immune response and lessening the duration and severity of a mild infection, while getting the lymph to pump and clear out some of the collateral damage.

Listen to your body and be prudent in your exercise decisions. But be honest about your condition. As a general rule of thumb, it is OK to exercise if you have a simple head cold or congestion (i.e. neck up) – in fact, it may improve the way you feel or, at worst, have little effect on duration or severity.[45] But if you don't feel well, especially if you have fever or body aches, stop daily exercise until you have recovered. If you have other medical conditions, such as diabetes or asthma, then you need to take those into consideration when making the decision, remembering it's always better to be safe than sorry.

IS EXERCISE THE 'BEST PILL' FOR CHRONIC DISEASE?

Absolutely yes. Any person who is capable of doing some light to moderate aerobic exercise should do so regularly (at least 30 minutes on three or more days per week) if they have a chronic disease. Exercise also helps to burn fat, improve the blood-lipid profile and prevent the development of excess visceral adiposity (belly fat) that results in inflamed adipose tissue and the subsequent development of insulin resistance that usually leads to type 2 diabetes. Short (10–15 minute) resistance exercise sessions three times per week can also help to delay sarcopenia. Exercise no doubt plays a role in reducing your risk of getting many chronic conditions. But what if you already have one? Living with a chronic disease really throws your body for a loop. Inflammation is sky high and energy levels are rock bottom. But exercise can be effective as a therapy for those who have developed metabolic or cardiovascular diseases, both to help reverse their disease and reduce the risk of other serious health complications.

Inflammation is one of the main targets of physical activity and, at the same time, it encourages immune regulatory cells that are lacking in many chronic diseases. Despite the advent of improved drugs, better outcomes are linked to better physical health. In the last few decades, a plethora of research has confirmed the beneficial role of physical activity as a tool to manage many inflammatory diseases. Just one session of moderate exercise can act as an anti-inflammatory therapy. Regular moderate activity can, in some situations, reverse symptoms and offer relief to those who suffer. For example, for heart disease, on which the NHS spends £17.38 billion per year (that's 18 per cent of all healthcare expenditure in the UK), data supports the notion of exercise as a 'first-line' therapeutic medical intervention. Findings like these have encouraging implications for other chronic diseases, since the best drug in the world couldn't have such a wide-reaching benefit.

So the most important thing is to keep moving, at whatever level you can. But you have to do it right:

- **Proper support.** First things first – check with your healthcare professional before exercising. And work with a fitness specialist if you are unsure what to do. Also, move to your own range and ability.
- **Don't overtrain.** Exercise can increase the inflammatory load and could cause exercise-induced leaky gut,[46] potentially exacerbating symptoms.
- **Budget your energy.** Keep a note of what activity you do. If you've overexerted, it will help you adjust accordingly. If you have an active autoimmune disorder you only have so much energy to spend in one day.

Cancer

Overall physical fitness is definitely a key quality that can help achieve good outcomes during and after cancer treatment. Research has shown that exercise is safe, possible and helpful for many people with cancer. But it's complex. When you think about how different we all are and how many types of cancer and treatment there are, it's difficult to write exercise guidelines to cover everyone. In general, at least 30 minutes a day, five days a week, of moderate-paced activity such as walking is encouraged, even during treatment. But exercise needs to be tailored to you, taking into account overall fitness, diagnosis and other factors that could affect safety. Check with your doctor before starting any type of exercise if you have cancer. Remember, being physically active doesn't have to mean joining a gym or an exercise class. It can also be walking to the shops, walking up the stairs, gardening or dancing.

Autoimmunity

An autoimmune disease doesn't make you fragile. You can still train, and evidence shows that you can probably go harder than you think, provided you allow for ample recovery and keep a lid on training volume. Some therapies for autoimmunity, such as low-dose naltrexone, even work in a similar way to exercise, inducing the body's feelgood endorphins, which bring pain relief. How you should exercise depends on the type of autoimmune condition you have.[47]

Exercise may benefit patients with rheumatoid arthritis and other inflammatory autoimmune disorders, such as lupus or ankylosing spondylitis, in a few ways. First, exercise could have potential long-term anti-inflammatory benefits related to effects on the signalling proteins that control inflammation. Second, physical activity stops other chronic symptoms from popping up, thus preventing a 'vicious cycle' of inflammation.

Painful inflammatory conditions like rheumatoid arthritis can make movement a daunting prospect, yet studies show that exercise consistently helps.[48] If your joints are compromised it could be worth working with a mobility specialist. People with multiple sclerosis really seem to benefit from exercise. Exercise even drives brain-derived neurotrophic factor (BDNF), which is reduced in MS. In the case of type 1 diabetes, exercise can play a role in reducing the amount of insulin needed because it upregulates insulin-independent glucose uptake by the muscles.[49] It's also safe, as long as your condition is well managed, the guidelines being that type 1 diabetics should monitor glucose levels before, during and after training to ensure the numbers don't get away from them.

Allergy and Asthma

You can't exercise away your allergies, but regular workouts can help manage symptoms, not only through the immune-regulating effects, but also by increasing blood flow which helps promote the removal of allergens from the body.

Exercise is good for asthma. In fact, many former world-class athletes have asthma, including runner Paula Radcliffe, cyclist Bradley Wiggins and footballer David Beckham. The key message is that as long as you're looking after your asthma well, and your symptoms are under control, you can enjoy any type of exercise, whether you choose to go for a brisk walk every day, join an exercise class or even sign up for a marathon. And by giving your lungs a regular workout you'll also cut your risk of asthma symptoms. But exercise alone can be a trigger for an attack in asthma sufferers. In fact, some asthmatics only get symptoms when they exercise, called 'exercise-induced asthma'.

Metabolic Disease

Physical activity is the most common intervention to tackle metabolic and associated problems like type 2 diabetes and heart disease. But you don't have to run a marathon to curb the symptoms. Exercise of any type reduces risk. Consistency clearly matters, but intensity is important too. Interval training improved all five disease markers of metabolic syndrome, and while very low amounts of activity were not enough to curb metabolic syndrome, moderate exercise – like a brisk walk – even in the absence of changes to your diet will do it. And for the people who stick to their sedentary lifestyles despite metabolic disorders, the prognosis generally looks worse.[50]

CHAPTER 7

Immunity-nourishing Nutrition

'Tell me what you eat,
and I will tell you what you are.'
JEAN ANTHELME BRILLAT-SAVARIN, 18th/19th-century French lawyer, politician and epicure

There is a reason why this chapter is last. I really wanted to elevate the status of our genetics, microbiome, sleep, stress, movement and ever-changing environment in nurturing strong immunity and encourage you to move you away from nutrition *in isolation* as a magic bullet to health and wellness.

As you have now learned from the previous chapters, the immune system is a huge constellation of different cells and molecules, each component of which not only has a unique function but an equally unique nutritional demand. Because of this complexity, the role of diet when it comes to the connections between nutrition, immunity and disease was, for a long time, poorly understood. We now have a much clearer picture, however. And it's not exactly what you might think. The relationship is not as linear as that between, say, smoking and cancer. In such a multifaceted system, nutrition can only ever be part of the picture. The link between diet, nutrients and immune responses is embedded in this intricate network of inputs, with our microbiota sitting at the interface of what we eat and how well our bodies can access and use those nutrients.

THE IMMUNITY DIET?

Before we start, let's be clear: there is no such thing as an immunity diet, per se. There is, however, a deeply entwined, yet highly complex relationship between nutrition and the immune system.

Some of you might have picked up this book hoping for a list of immune-nourishing nutrients. Maybe you thought that through supplements, superfoods and vitamins your immunity will become stronger, boosted even. In this chapter, as we consider how eating can support your immunity, it's important to note that there are many claims about foods that supposedly *enhance* immunity, but most are not supported by good evidence.

You might be aware that vitamin deficiencies are bad for immunity. What might not be so obvious, though, is that the immune system suffers in both conditions of under- *and* over-nutrition; too much food can be bad as can too little. Ultimately, we want a balanced immune system, and this requires a balanced diet. But there is so much confusion over what that looks like. There are macros, micros, vitamins, minerals, antioxidants – and the list goes on. But what exactly are they? And what do they have to do with eating for immunity?

THE FOOD PRISON

Hippocrates' philosophy – 'Let food be thy medicine and let medicine be thy food' – is very popular, showing up on countless social-media posts. And therein lies a certain optimism: if you're sick, just change your diet to get better. But perhaps we are failing to see the real issue. Food is not medicine. The idea of healing what ails you with food is tempting – medicine can be scary, have horrible side effects and may not always treat the cause, only managing symptoms. But medicine *can* work. And while good nutrition can absolutely support our health from cradle to grave, when we turn food into medicine and cultivate an 'eat-to-live-healthily' mentality, we risk stripping it of everything else – we erase histories, celebrations and memories. I call it the food prison: the dangerous, damaging, yet completely socially acceptable facet of modern food culture; an imprisoning by a range of self-imposed dietary rules, excessive food restrictions or extreme opinions where food views are almost an identity. And the stress of eating food devoid of joy and social connection – 'eating clean', as if your life depends on it – is definitely not helpful for immunity.

Food is much more than nutrients. It fulfils a profoundly social urge and is something that is almost always shared: people eat together, mealtimes are events when the whole family or friends come together. Food is also an occasion for celebrating, for distributing and giving, for the expression of altruism, whether from parents to children, family to friends, visitors or strangers. Today we have more knowledge about nutrition than ever before, but possibly the best thing we can do with that information is use it to lead us back to the dinner table and to start enjoying food again.

Having said all that, both common sense and modern medicine tell us that food choices constitute one of the greatest opportunities to improve your immunity and, by extension, your wellbeing. So, let's have a look at some food facts.

THE MICROS

Science often uses complex terms to describe beneficial food components. Micronutrients are molecules like vitamins and minerals, which we need in small quantities, yet are essential for our health. Science now recognises around 13 essential vitamins (that we know of), as well as some 20 minerals that must be consumed to keep our bodies in good running order. Vitamins and minerals are vital constituents of our diets that are crucial for health, and a lack of any of them can cause diseases of deficiency. Immunity requires all the micros for optimal day-to-day functioning. So it follows that having a specific deficiency in any of these could leave your immune system compromised. I'm not going to go through all the micronutrients, but just pick out a few useful pointers for strong immunity.

Deficiency diseases like scurvy (vitamin C), beriberi (vitamin B1), rickets (vitamin D) and pellagra (vitamin B3) are rare in today's modern Western world. They often result from cultural and economic disadvantage or just plain ignorance, but many years of research have allowed the majority of us to overcome or avoid them. The study of such conditions, however, has taught us the important influence of these micronutrients on immunity – we need of all them to be able to defend ourselves from infection and keep our immune systems running smoothly.

Vitamin C

I'm sure when it comes to vitamins and immunity, most people envisage popping vitamin C tablets. But the good news is that for most of us vitamin C is a practically unavoidable component of a normal balanced diet and we can obtain our RDA (recommended dietary allowance) easily through consumption of fresh fruit and vegetables.

In the 1960s and 70s, Nobel prize winner Linus Pauling promoted the idea that vitamin C is a cold cure-all, recommending megadoses of vitamin C, thousands of times higher than recommended in the UK today.[1] Since then, multiple randomised controlled studies have examined the effects of the vitamin on the common cold and the results have been fairly disappointing.[2] It's true that vitamin C plays a key role in immunity. But although a deficiency can lead to higher susceptibility to infections, supplementing does not reduce the risk of catching a cold. If you do catch a seasonal lurgy, vitamin C supplements of 1–2g per day have several benefits, including reducing the symptoms and severity, decreasing recovery time by 8 per cent in adults and 14 per cent in children, on average. Vitamin C appears to have even stronger effects in people who are under intense physical stress. In marathon runners and skiers, vitamin C supplementation almost halved the duration of the common cold.

Vitamins A and D

While vitamin C is, of course, important, for immunity vitamins A and D take centre stage (see pp. 147–9 for more on vitamin D). Vitamin A is a micronutrient that is crucial for maintaining vision, promoting growth and development. Despite the fact that it plays a critical role in our immunity, it's often misunderstood. It helps regulate the immune system and protects from infections by keeping skin and tissues in the mouth, stomach, intestines and respiratory system healthy. The confusion is perhaps because, like so much in nutrition, context matters: vitamin A is key for the immune system to remain tolerant and anti-inflammatory, particularly in the gut. There has been some work to trial vitamin A as a therapy for inflammatory diseases, particularly in the gut (such as inflammatory bowel disease). Unfortunately, this has not been found to be beneficial and high doses of vitamin A can even turn up the heat and promote inflammation.

The amount of vitamin A we should get daily is a bit complicated because it depends on where it comes from and your age. Vitamin A from plant-based sources is called carotenoids, and from animal sources it's called retinol. Retinol is more bioavailable than the plant source, since the body has to convert carotenoids to retinol. Luckily, deficiency in vitamin A is not something we often deal with in developed nations and it is difficult to overdose through diet. You can get this immune-nourishing vitamin from sweet potatoes, carrots, kale, spinach, red peppers, apricots, eggs or foods labelled 'vitamin A fortified', such as milk or cereal. If you do take a supplement that contains vitamin A, just be careful with the dose, especially if it is in the highly absorbable retinol form.

Zinc

Zinc is an essential mineral that's found in almost every cell and plays a multitude of roles in the body. It is critical for the normal development and function of cells that mediate both innate and adaptive immunity (see pp. 7–11).

Because zinc is not stored in the body, regular dietary intake of the mineral is important in maintaining the integrity of the immune system. Our ability to absorb zinc seems to decline with age, making elderly people particularly at risk, and a deficiency has been shown to impair defence against infection.[3] Zinc appears to reduce the ability of cold viruses to grow on or bind to the lining of the nose, and improves the power of special immune cells to fight infection, suggesting there is a small benefit in supplementing with zinc during the winter months for prevention of infection. There is also some evidence that starting to supplement with zinc upon the onset of symptoms (when the need for extra zinc increases) may help to reduce the severity and duration of an infection.[4] Zinc lozenges in particular seem to be effective in the short term, but there's still debate as to the best formulation and dose.

Long-term use – more than six weeks – can lead to copper deficiency and an irritated digestive tract.

Selenium

Selenium is another essential micronutrient with antioxidant properties that is involved in many biological processes throughout the body, including immune responses. It is not surprising, then, that getting adequate dietary selenium strongly influences how well we deal with infections.

Selenium is found in the soil and naturally appears in produce grown in the earth. It's quite rare to be deficient, unless you are in an area that has a known selenium soil deficiency. Studies looking at selenium status on immunity have suggested that, in the absence of an overt deficiency, more selenium might not always be better and could even drive unruly inflammatory processes in the body.

Should You Supplement?

Despite most people believing, as the industry tells them, that supplements will make them healthier, there is little to no evidence that this is the case. Vitamin and mineral pills are a convenient way to plug specific gaps in your diet and they protect from diseases of deficiency. But if you are not deficient, there is negligible evidence that they will make the immune system work better than it already does at baseline and they are *not* going to be effective at mitigating risk of infection.

When it comes to nutrition, I advocate a food-first approach, with a few exceptions. Why? I've long had a hunch that we absorb nutrients more effectively from food than from dietary supplements. It makes sense on an evolutionary level, since we've evolved with our gut microbiota to digest and extract nutrition from the food that's available. Plus, food has some unique nutritional qualities that are vital for immunity and which we cannot get from supplements alone – for example, fibre. A recent study confirms my hunch, showing that dietary supplements do not provide any benefits when looking at mortality rates. Micronutrient benefits occurred only when consumed in real food form. And most concerning: they found that excess intake of some nutrients may cause adverse effects and increase risk of death from disease. But it is practically impossible to consume too many nutrients from real food.

Given the evidence for the lack of health benefits for supplements in the majority of the healthy adult population, why do over 40 per cent of us regularly use them? The immensely beneficial effects in preventing the deficiency diseases of times gone by when overt nutritional deficits were common have given supplements the halo of a magical panacea. Nowadays, for some, they are an insurance policy (or simply an excuse) against a less-than-perfect diet. Others take them because they can't – or won't – eat certain foods. And not to forget the biology of belief: the very real placebo effect of pill-popping. Then, add to this the media and the relentless marketing of vitamins, herbs and other supplements, with vague claims such as 'maintains immune function' – all legal, so long as they don't specifically say that they actually treat or cure anything. Remember, supplements tend to be much less regulated than prescription 'drugs' and can usually be purchased over the counter or online – but this doesn't mean they're always safe. In fact, it is worth scrutinising any supplement you are considering because they are so unregulated.

When might supplementing be necessary?

The primary reason for taking a supplement should be to prevent deficiencies or a lack of essential vitamins and minerals. If your diet in general is less than perfect, then improving what and how you eat is where you want to start. No number of pills will negate bad food intake. But there may be certain circumstances in which you need to supplement.

Extra zinc and vitamin C may be useful for hard-training athletes. According to the UK's 2018 National Diet and Nutrition Survey,[5] magnesium is something we are all a little deficient in. There are some nutrients it's almost impossible to get in adequate quantities from food alone – vitamin D in particular, and especially in the UK and countries at similar latitude where, during the winter, the sun isn't strong enough to allow your body to produce it naturally. If you have, or are at risk of, osteoporosis, your doctor will likely recommend calcium and vitamin D supplements. But vitamin D *only works well* if you also have vitamin K2, approximately half of which is produced by our microbiome. So already you start to see a complex picture emerging in which a multi-pronged approach is needed to take care of those gut bugs to help our vitamin D function properly.

Anyone over the age of 50 may need a vitamin B12 supplement (because this nutrient becomes harder to absorb from food as we age), as do those following a vegetarian or vegan lifestyle. Older people tend to have smaller amounts and often have less variety in their diets, and they also might absorb nutrients poorly. Micronutrient requirements have been noted to change over the course of life too.[6] So dietary supplements may help older people maintain a healthier immune system.

Triage Theory of Micronutrients

Our bodies naturally 'triage' on a daily basis. Cells sacrifice nutrients from non-survival functions for immediate physiological needs – say, diverting nutrients from tissue repair to meet a more critical need. So a nutrient-poor diet may leave you on the fringes of the RDA for vitamins and minerals. You might not have an overt deficiency, but nutrients will be diverted away from preserving longevity to make sure you are meeting more immediate and critical needs. This sets the stage for poor long-term health and is emerging as a key driver of lifestyle-related diseases.

Although much more research is needed in this area, it seems that the requirements for short-term health are different from those for longer-term health. This might be particularly important for the elderly, who may need more of certain vitamins and minerals compared with younger people, or those suffering chronic diseases, especially if their overall diet quality is poor. In these cases, it could be that a multivitamin supplement plays a role as an insurance policy in addition to urging people to eat a more balanced diet. But while some may be good, more is not better – make sure you don't exceed what you need.

PHYTONUTRIENTS – THE POWER OF PLANTS

While micronutrients are essential for health, phytonutrients are very beneficial. Phytonutrients are biologically active chemical compounds found in plants. They act as a natural pesticide, helping to protect plants from predators. So it's no surprise that regular consumption helps to prevent us from getting sick too.

Currently, phytonutrients are referred to as non-nutritive, meaning we don't have a specific recommended daily intake or reference amount deemed necessary for health – partly because they are not as essential to survival as vitamins and minerals. But there are many ways in which eating a variety of phytonutrients provides us with extra 'immune-nourishing' benefits, not only protecting us from infection but warding off long-term chronic disease – not least through their unique antioxidant, anti-inflammatory and antimicrobial properties.

Plus, they are key to unlocking our own internal antioxidants that protect our delicate cellular machinery, reduce the burden of oxidative stress and remove those 'zombie' immune cells (see p. 84). Regular consumption of phytonutrients from plant-based foods reduces the risk of infections and certain illnesses, such as cardiovascular disease and some cancers.

There are many different groups of phytonutrients. Here are some examples:

- **Flavonoids**, such as anthocyanins and quercetin found in soybeans, onions, apples, tea and coffee. Anthocyanins have been shown to be great pain relievers with anti-inflammatory properties.
- **Carotenoids** found in red, dark green and orange plants such as tomatoes, carrots, sweet potatoes, watermelon, leafy greens. Perhaps the best-known is lycopene (in tomatoes), which has long been associated with anti-cancer properties. While all tomatoes contain lycopene, the skin has the highest concentration and cooking is the best way to convert the lycopene into the most absorbable form. Lycopene is fat-soluble, so if you cook tomatoes with a little olive oil, the amount absorbed goes up three-fold. The San Marzano heirloom variety, from southern Italy, has one of the highest levels.
- **Carnosol**, a bioactive compound extracted from Mediterranean herbs such as rosemary and sage, has been found to have promising anti-cancer and anti-inflammatory properties.
- **Resveratrol** is produced by plants as a natural fungicide. It is found in grapes, red wine and berries.
- **Epigallocatechin-3-gallate (EGCG)** is an immune-nourishing polyphenol with natural anti-inflammatory and antioxidant properties found in quantities 16 times higher in green tea than black tea.

- **Sulforaphane and diindolylmethane (DIM)** are among the favourable compounds in cruciferous vegetables like broccoli, known to nourish our immunity. Remarkably, broccoli sprouts contain up to 100 times more sulforaphane than the full-grown vegetable. Sulforaphane activates a special genetic pathway in our cells known as Nrf2 more potently than any other known naturally occurring dietary compound. This gene, a master regulator, controls over 200 other genes, turning off inflammation and upping our own internal antioxidants. In a sense, we're talking about an 'On' switch for some of our native stress responses, making our bodies more resilient.

There are over 25,000 different phytonutrients recorded across many foods – not only fruits and veggies, but also beans and pulses, tea, coffee, red wine, cacao, herbs, spices, condiments and olive oil – and their immune-nourishing effects cannot be explained by selecting just one or two. The various types are metabolised differently in the body and may produce diverse health effects.

Using the power of plants to nurture our immunity is often considered 'unconventional' in mainstream medicine, and controlled clinical studies are still quite small. It's important to note that although there may be a vast body of research on their health effects, many phytonutrients have only been looked at in test tubes. On a practical level, most of them are found in such small amounts that it requires a huge intake to match the amounts used in most research studies. That's not to say we shouldn't prioritise them, but we should look beyond seeking out a health benefit from an individual phytonutrient. It's their collective power that we are after, and as a rule of thumb, more colourful fruit and veggies equals more phytonutrients.

The phytonutrient content of plants varies depending on the conditions in which they are grown. For example, plants make more when there are 'stresses' in their environment – for example, when it's cold.[7] Phytonutrient content tends to be higher when they are grown organically without any 'help' from pesticides.[8] This is why seasonal eating is so important for our health, something that traditional cultures have embraced and now science is catching up on. This is a field of research called 'chrononutrition', which focuses on the interaction between biological rhythms, nutrition and metabolism.

A Word of Caution on Phytonutrient Supplements

There have been many attempts by supplement companies to capture the properties of phytonutrients in pill form, but this is problematic, unregulated and unlikely to have the same benefits as consuming phytonutrients in a meal. In some cases, they might even be detrimental to our health. The dose makes the poison: phytonutrient supplements, often advertised as antioxidants, tend to contain doses far higher than would naturally be consumed through diet. Beta-carotene pills, for example, may actually increase cancer risk, as opposed to the whole carrot, which may lower it. Taking high levels of antioxidants in supplement form rather than food also blunts the health benefits of exercise.[9,10]

THE MACROS

Macronutrients are molecules that include carbohydrates, proteins and fats. These three all have their own specific roles and functions in our immunity, and we need them in large quantities. While protein is necessary to generate the building blocks of our bodies, fats and carbs provide us with energy.

Pumped on Protein

When people talk protein, normally muscle building gets all the attention. However, your immunity needs protein to run on all cylinders, especially when you are busy. In the body, amino acids (the building blocks of protein) play critical roles, maintaining muscle, building blood cells, growing your hair and forming enzymes like haemoglobin to shuttle oxygen through your bloodstream. And protein is vital to your immunity too, building and repairing body tissue and fighting infections. Immune-system powerhouses such as antibodies, cytokines and other immune-cell responses are built from amino acids, so they rely on an adequate intake of protein. Lack of protein can really compromise immunity, particularly for our T cells – the master regulators of both responding and regulating inflammation. Animal studies have shown that the immune system can be significantly compromised with even a 25 per cent reduction in adequate protein intake.

Proteins are molecules made up of long chains of amino acids of which there are 20 different types, and whose sequence determines a protein's structure and function. Nine of the amino acids are essential, meaning they must be provided through the diet, while the body can easily make the other 11 (non-essential) itself, though some are conditionally essential – that is they become essential under certain circumstances, usually when the body is stressed or sick. A complete or whole protein is a food source that contains an adequate proportion of each of the nine essential amino acids. Animal protein sources are complete; plant foods are incomplete. However, you can combine various plant foods to create a complete protein, and this has intuitively been the case for many cultures around the world – for example, beans and rice in Mexico, bean stews with bread in Africa, pasta with beans in Italy and lentils with rice in India.

Some of the amino acids are particularly important for immune functioning, especially arginine. Although classified as a nutritionally non-essential amino acid, arginine is the physiological substrate for the synthesis of nitric oxide (NO), which is a key mediator of immune responses. Much experimental and clinical data support the notion that arginine is an essential nutrient for both innate and adaptive immune systems in humans and other animals.[11] Found in nuts and seeds, meat, legumes and seaweed, it promotes T cell development, growth and thymic integrity. Glutamine (see p. 100) is also important. When we are struck with an infection and huge armies of immune cells are suddenly mobilised to protect us, the immune system's nutrient needs dramatically increase. During these times, it uses glutamine. Being sick or having a chronic inflammatory disease can deplete our internal stores, and we need to take care and replenish. That's why intravenous glutamine and arginine are often given to people who are seriously ill to support their immunity, and as a protective therapy in pre-surgery patients.[12] That's not to say you need to supplement. But know that your needs may vary depending on your health status. The daily recommended protein intake is 0.8g per kilogram of body weight. A rough guide would be around 50g per day for an average person. To give you an idea of what that looks like: a cup of quinoa is 8g, a cup of lentils is a whopping 18g. An egg will add 6g. And a 115g piece of chicken is another 28g. Spinach, edamame, beans, peas, broccoli and tofu all readily offer additional sources. Athletes or those looking to build muscle may want to aim for more. Aiming higher with your protein can also make a difference if you're extremely physically active, have a stressful life load, are struggling with a scratchy throat or just feeling run down. Protein is also a great satiating nutrient, which can help with hunger when cutting calories if you are trying to lose weight. Advancing age demands greater protein intake, with extra protein helping to maintain muscle mass (so important for strong immunity) as we get older. Those over age 65 should aim for at least 1.2g of protein per kilogram of body weight and for older weightlifters 1.5g/kg of body weight each day.

Protein shakes may be a convenient way to get protein if you are short on time. For the most part, though, you need to look to your diet to meet daily protein needs. Research suggests that if you are healthy and getting enough calories in each day, you will also be consuming adequate protein. Getting protein from foods rather than a shake has the benefit that other essential nutrients and fibre will also be provided. Plus, who wants to drink a meal? Aim to source from as much variety as possible. It is not just a deficiency of specific amino acids that can compromise the immune system, but an imbalance among all the protein building blocks. To ensure you have them all covered, aim to include diverse sources. Relying on plant-based sources alone can be challenging since you'll need to eat a much larger quantity compared with animal sources. This can also come with a lot of extra calories. Try things like soy, tofu, textured vegetable protein and seitan to reduce the food volume. In particular, soy and quinoa are considered a higher-quality protein source because of their amino-acid profile.

FATS: THE GOOD, THE BAD AND THE IN BETWEEN

Despite their controversial past, fats are essential for health. They serve our body in three main ways: they provide a great source of energy, they're the building blocks of our cells and the precursor for signalling molecules that regulate our immunity. Dietary fat also brings with it the important fat-soluble vitamins A, D, E and K.

We may think of fat as being just one thing, but not all fats are created equal and it's the composition of the different types of fats in the diet that matters most. Fats found in foods are typically saturated or unsaturated. Most foods usually contain a mixture of both, so you may hear foods being described as high or low in one or the other.

Saturated fats are generally solid at room temperature – common sources include red meat, whole milk and other whole-milk dairy foods, cheese and coconut oil. Diets *high* in saturated fat are generally bad for health, as these fats can 'short-circuit' immune cells, acting as danger signals, a molecular stress on the body, triggering a form of oxidative stress. This sends a message to the immune-signalling centre called the 'inflammasome' to produce an inappropriate and damaging inflammatory response. So the emphasis should be on foods higher in unsaturated than saturated fats as part of a healthy, balanced diet. Dairy may be the exception, as the evidence is shaping up that the saturated fat in dairy is not as harmful as in other sources. But this doesn't apply to the dairy found in desserts and butter – rather cheeses, yoghurt and milk.

Unsaturated fats come in different types: trans fats, monounsaturated and polyunsaturated fats.

Trans Fats

Naturally occurring trans fats are found in small amounts in dairy products (for example, cheese and cream) and also beef, lamb and mutton, and products made from these foods. Trans fats may also be produced when ordinary vegetable oils are heated to fry foods at very high temperatures, which is one reason why processed and convenience foods can sometimes be high in trans fats. Foods that are produced from or use hardened vegetable oils typically contain some trans fats (biscuits, pies, cakes and fried foods). Fat spreads and margarines that have hydrogenated vegetable oil as an ingredient will usually contain some trans fats. Until the 1980s, margarines contained 10–20 per cent trans fats, but modern versions are much lower or virtually trans-fat-free. No matter their dietary source, trans fats are almost always bad for us and have been found to alter our immune response for the worse. But consumed at low levels, they are unlikely to have a significantly harmful effect.

Monounsaturated Fats

Monounsaturated fats are found in plant foods like avocados, nuts and cooking oils like olive oil, but also animal meats and produce. Science supports the idea that monounsaturated oils like olive oil are healthful, and capable of benefiting immunity and reducing unnecessary inflammation. But accounting for the source is important, given that monounsaturated fats from animal products come with a higher intake of less healthful saturated fat, whereas in plant sources they come conveniently packaged with fibre, phytonutrients and micronutrients. The discovery that monounsaturated fat could be healthful came during the 1960s.

Polyunsaturated Fats

Polyunsaturated fatty acids include omega-3 and omega-6. These are quite different from most other fats when it comes to immunity. Not only are they important components of our cell membranes, affecting how fluid they are, they are biologically active, acting as a launch pad for making specialised immune-signalling molecules that can either promote or fight inflammation. Omega-3s are used by our immunity to make special anti-inflammatory and pro-resolving mediators, playing a role in taming inflammation and suppressing key inflammatory genes. They are found in olive oil, flaxseed oil and fatty fish such as salmon, sardines and mackerel. Omega-6 fats are the raw materials for inflammation. The most common omega-6 fatty acid is linoleic acid (LA), which is metabolised to arachidonic acid (AA) – a pro-inflammatory inflammatory-signalling molecule. It is found in our diets in cooking oils, poultry, and some nuts and seeds.

Omega-3 science is not straightforward: alpha-linolenic acid (ALA – another omega-3), DHA and EPA have some potentially important differences.

- **ALA** is the 'parent' one, found mostly in plant foods, particularly flaxseeds, chia seeds and walnuts. It is considered the essential omega-3 since it cannot be synthesised by us humans. It can be converted to DHA and EPA in the body via a series of enzymatic reactions, but this is relatively inefficient. Ethnic distribution of genetic variations appears to have a huge impact on how efficient the conversion is.
- **EPA and DHA** – we can make these (albeit not very well). They are considered conditionally essential nutrients, and so consuming them directly from foods and/or dietary supplements is a practical way to increase levels of omega-3s. Fish provides the most available source of EPA and DHA.

Omega-3:Omega-6 ratio?

Nutritionally, omega-3 and 6 are essential fats – meaning we need to get them from our diet as our bodies can't synthesise them. If we don't get them from our diet in adequate amounts, then we are at risk of deficiency. Omega-3s share the same enzymatic pathways as omega-6s, meaning they are in competition. One way to look at it is omega-3 and omega-6 having opposing functions in the body – like a switch between pro- and anti-inflammatory. At first it might seem like omega-3 are good and omega-6 are bad, but this is an oversimplified view.

Ultimately, both omega-3 and omega-6 are important for health, and should replace saturated and trans fats where possible. While saturated fats are the most inflammatory, overabundance of omega-6 polyunsaturated fats, such as those found in most cooking oils, are also inflammatory (although this has only proved true in test-tube and animal studies, and has so far failed to accurately reflect the complexity of human physiology). While it's true that our modern-day diets tend to be lower in omega-3 and higher in omega-6, the optimal ratio – if any – has not been well defined.

Omega-3 and 6: fish, brains or supplements?

Oily fish (especially fish roe) provides the most available source of omega-3. But omega-3-rich oils and supplements are now being touted as natural wonder drugs. Should you take them, though? The answer, like so many things in science, is: it depends. There is no absolute rule about how much omega-3 a person needs. Nor is there an established upper limit (just be aware that very high doses of omega-3 fatty acids in healthy individuals could decrease the potential of the immune system to destroy infections). Ultimately, different groups of people may need different amounts, so you should discuss this with your doctor or nutrition professional, but here are some pointers:

- **Do you regularly consume oily fish?** Even with no apparent deficiency signs or health conditions, a diet rich in omega-3s plays a key role in overall health and wellbeing. Fish is a great source of EPA (eicosapentaenoic acid) and DHA (docosahexaenoic acid) – both omega-3s, found in the brains of land mammals (omega-3 is not readily accessible from their muscle tissue), leading some to suggest that our inland forebears ate the fat-rich brains of land animals. The form found in fish (in particular, fish roe, as mentioned above), known as the triglyceride form, is more bioavailable and preferentially imported into the brain. If you don't often eat oily fish (or brains), supplementing with omega-3s is important.

- **Do you consume too much omega-6?** Eating a diet rich in omega-6s doesn't lead to inflammation in itself. But overconsumption of them in the absence of omega-3s can be problematic, particularly if you have a chronic inflammatory condition. Be mindful of the source of omega-6 too. Modern processed foods also tend to contain higher saturated fat, and less fibre and phytonutrients, which we know isn't good for health. On a population level we have increased omega-6 consumption in the last 100 years due to the development of technology at the turn of the century that marked the beginning of the modern vegetable-oil industry.
- **Do you have an omega-3 deficiency?** Clinical signs of omega-3 deficiency include a dry, scaly rash, decreased growth in infants and children, increased susceptibility to infection and poor wound healing, hair loss and problems conceiving.
- **Do you have a chronic inflammatory disease?** Some of the most interesting research on omega-3 supplements comes from treating the symptoms of chronic inflammatory diseases. Omega-3s are helpful for symptoms of inflammatory bowel disease, improving tender joints in rheumatoid arthritis and symptomatic improvements in psoriasis. New research also suggests that omega-3s can cross into the brain and may help lower the inflammation that contributes to Alzheimer's

disease. Omega-3s in foods may reduce symptoms of childhood asthma, while omega-6s may aggravate them.[13]

Keep in mind the crucial role of our microbiome in helping us use these important dietary components. It was recently shown that one of the beneficial effects of omega-3 fish-oil supplements relies on gut bacteria producing anti-inflammatory postbiotics. But not everyone can produce this chemical from their mix of microbes, explaining why these supplements don't always work (and perhaps why some of the clinical trials have given conflicting results).

Here again, a food-first approach is best – two portions of oily fish per week is recommended, or algae supplements if you prefer not to eat fish. But the benefits of eating fish extend beyond those of omega-3; it's associated with reduced risk of heart disease in both epidemiology and clinical trials, and appears to help curb the risk of depression.

Don't Mention the C-Word

In the past five years, the reputation of carbohydrates has swung wildly. They've been touted as the feared food in fad diets, while being associated with lower risk of chronic disease. So which is it? Good or bad? The short answer is: they are both.

Carbohydrates are chains of small, simple sugars that are broken down and absorbed by the body as glucose – an energy substrate for all the cells in the body. They can be simple monosaccharide (one sugar molecule, i.e. glucose) or disaccharide (two sugar molecules, e.g. sucrose – table sugar). Complex carbohydrates are multiple simple sugars connected together. All carbs are eventually broken down into glucose.

When you eat a food containing carbohydrates, the digestive system breaks down the digestible ones into sugar, which enters the blood. As blood-sugar levels rise, the hormone insulin is released, prompting cells to absorb blood sugar for energy or storage. As this happens, sugar levels in the bloodstream begin to fall. Some more complex carbohydrates take longer to break down and be absorbed (leafy greens, sweet potatoes, beans, whole grains) and tend to contain other beneficial nutrients like fibre, vitamins and minerals. Most simple carbohydrates tend to lack nutrients while being high-calorie and absorbed quickly (think sweets, white bread, juice).

Hi-carb, low-carb or no carb?

Although carbs are controversial, they play a massive role in keeping your immunity robust. In Chapter 6, I shared the important role played by carbs in fuelling up for optimal exercise performance. Fighting infection is a costly affair and immune cells have high demands for glucose. If levels are too low for too long, it can dampen the potential for your immune system to respond to an infection, and can reduce the number and function of key immune cells like neutrophils, NK cells, and B and T cells.[14]

However, we know that excess carbohydrate nutrition is related to oxidative stress and inflammation, and that blood-sugar control is a cornerstone of a long and healthy lifespan and fighting off chronic disease. If blood-sugar levels remain raised for too long, it makes our immune cells less able to do their job, leaving us open to infection and cancer. This is because the high glucose levels unleash destructive molecules that cause the formation of harmful advanced glycation end products (AGEs) that can interfere with how the immune system works. Rising blood sugar is normal after eating. It's known as a blood-glucose excursion when your blood glucose spikes above its normal level. This triggers insulin in healthy people, which works to lower blood glucose back to baseline. This also helps to rein in inflammation. We know from studies of diabetic patients (who have high amounts of sugar in the blood that their bodies struggle to remove) that this increases the risk of dangerous infections and numerous inflammatory health complications. But when it comes to assessing the risk of complications, it's not just how high your levels go – it's also how much variation in blood-sugar levels you're experiencing.

Fortunately, you can get the benefits of carbs without worrying too much. The easiest way to do this is to eat at regular intervals when you feel hungry, and choose more good-quality carbs (whole grains, beans, vegetables and fruit), which are full of fibre, phytonutrients, vitamins and minerals, and are absorbed slowly, thus avoiding sugar spikes. And you can minimise the health risks of carbs by eating fewer of the refined and processed ones found in cakes and many baked foods that have been stripped of beneficial fibre. A carb:fibre ratio is a useful rule of thumb for identifying a whole-grain food. A product should have at least 1g of fibre to every 10g of total carbohydrate. It is an easy equation to calculate if you look at the nutritional information on a food label: simply divide the total carbohydrate in grams by fibre (in grams).

It's all too easy to overeat low-quality refined carbs. And as I mentioned earlier, maintaining a healthy weight is also important to avoid unruly inflammation.

Healthy people found that consuming 40g of added sugar from just one can of sugar-sweetened fizzy drink per day led to an increase in inflammatory markers.[15] These people tended to gain more weight too. The reality is, the relationship between carbs and health might have to do with what we are *not* eating – dietary fibre (of all kinds).[16] Context matters, though, and no nutrient is truly bad. But we now know that swapping out 'bad' saturated fats and replacing them with refined carbs is bad for our health. Rather, we should aim to reduce these unhelpful fats, replacing them with unsaturated ones.

GLUTEN-FREE – FIX OR FAD?

Gluten is a family of hundreds of proteins – notably, glutenin and gliadin – found in grains like wheat, rye and barley. Mixed with water, they form a glue-like consistency, which makes dough elastic and helps bread rise. 'GF' (gluten-free) labels are commonplace in restaurants and grocery aisles, and recent polls suggest that one in five of us healthy adults intentionally limits or restricts gluten intake. But is 'gluten sensitivity' real or yet another food fad?

Around 1 per cent of us in the UK have coeliac disease, an autoimmune condition that occurs in genetically sensitive individuals. Remember those immune compatibility genes? Well, specific types (known as DQ2 or DQ8 heterodimers) mean the immune system has a different interaction with gluten peptides – and not in a good way. Activating your immune army, led by T cells unleashing a host of inflammatory cytokines, the immune defences start to work on damaging the gut mucosa. Eating gluten if you are coeliac causes the lining of the small intestine to become damaged, leading to diarrhoea, bloating, fatigue, anaemia and poor nutrient absorption. Other parts of the body may be affected too.

Gold-standard diagnostics for coeliac disease include blood testing and confirmation with a gut biopsy to check the level of any inflammation or damage to the gut mucosa. Coeliac disease is a lifelong condition and a gluten-free diet is the only treatment for it. If gluten is introduced back into the diet at a later date, the immune system will react and the gut lining will become damaged again. Even eating a tiny crumb will cause this horrible damage to the gut wall and can lead to an elevated risk for cancer.

Is there any reason not to be gluten-free?

For the millions officially diagnosed with coeliac disease (like myself), going gluten-free for life is essential. Aside from coeliac disease, though, many people claim to experience health problems relating to gluten. Almost 20 years ago, the existence of 'non-coeliac gluten sensitivity' was first proposed. But since then, follow-up studies have reached the opposite conclusion: no evidence of a specific response to gluten. Gluten wasn't the culprit; the cause of digestive upset in these studies was psychological – a 'nocebo' effect (the belief that when we consume something it's going to mess us up).

Nevertheless, the suggestion of an effect of gluten outside the classic association with coeliac disease or classic wheat allergy remains a topic of significant speculation. Most up-to-date accumulating evidence points to its existence in an extremely small number of people – far fewer than the millions buying gluten-free products. Unfortunately, we currently have no valid test or biomarker for non-coeliac gluten sensitivity and there is little consensus in the science.

But is there any risk in removing gluten? For many people, 'gluten sensitivity' is actually a response to other components in gluten-containing products – like fructans in bread, which can cause bloating when they ferment in the gut, or food additives and preservatives. But coincidentally, some of the biggest sources of fructans are removed when adopting a gluten-free diet, leading people to conclude that gluten-free 'works'.[17,18,19]

If you suspect a gluten intolerance, it might be more to do with your mix of gut microbes struggling to digest the wheat. In which case, working on your gut health might be a better place to start than a total exclusion of gluten, which might make things worse. Then it really depends *how* you go gluten-free – what alternative foods are you going to choose? Are you going to eat a diet that is gluten-free by nature, choosing naturally gluten-free grains, fruits and vegetables? Or swap out gluten-containing products for poor-quality refined and processed gluten-free cakes, cookies and bread?

Going gluten-free may even cause health issues. The implications of adopting a gluten-free diet in terms of nutritional deficiency has, as yet, received little consideration, but gluten-free cereal products contain fewer essential micronutrients and are lower in fibre. In fact, recent studies have shown than healthy adults going gluten-free had substantial changes to their microbiomes and their immune cells were worse at responding to infection. I wrote earlier about the stress of trapping yourself in a 'food prison'. Well, perhaps considering that true non-coeliac gluten sensitivity is quite rare, advising a gluten-free diet for all might lead to more stress than it's potentially avoiding. So next time you come across a 'gluten-free' label, think twice – unless you are a coeliac, of course.

FEEDING YOUR IMMUNE SYSTEM – THE IMPACT OF METABOLISM

Metabolism is the set of life-sustaining chemical reactions that help convert the macronutrients in food to energy. But food intake is sporadic (usually in three major meals each day), while the way we expend our energy is continuous, with variations that have nothing in common with when we eat. The body has developed complex systems that direct excess macronutrients into storage for later use. And since immunity relies on metabolism to fuel its activities, it has developed quite sophisticated ways of detecting the availability of each of these macronutrients too.

Immunometabolism – that is, how our metabolism is integrated with the fate and function of our immune cells – is a swiftly growing area of research. In healthy people, the immune system is normally in a 'quiescent' state – it's on the lookout, but not actively in fight mode. When it is time to attack, immune cells start to increase their metabolism, absorbing more nutrients in order to produce the proteins essential for defence, such as cytokines and antibodies.

Scientists have long recognised that problems with metabolism have links to inflammation, while some inflammatory conditions, such as rheumatoid arthritis and other autoimmune diseases, have a metabolic component. This means that too much or too little of one or another macronutrient can have quite a profound knock-on impact on immunity. To date, there are very few studies examining the impact of meals with different macronutrient composition on immunity. But we are starting to get a few clues. We know that immune cells can switch their metabolic fuel source between fats and carbs, depending on what they are doing. When the body is combating a cold or running a fever, it needs energy to build an army that will fight off infection and recover. This upregulates their need for glucose – the breakdown product of carbs. In fact, inflammation actively promotes the availability of glucose. Fat, on the other hand, is the preferred fuel for the slow burn, keeping our immune cells ticking over when not actively deployed in battle. It stands to reason that the ability to sense which fuel sources are available is part of our immunity's decision-making process when choosing whether to mount an inflammatory response and to what degree.

The problem is that this is such a new field that experts are still trying to work out how to alter diet in order to switch off an unwarranted inflammatory response. Can we change disease course by altering macros? Or by not providing any of a specific macro at all?

Feed a Cold, Starve a Fever

Centuries ago it was believed that colds were brought about by a drop in body temperature, and that by eating more we could help raise it and thus kick the cold. The 'starve a fever' recommendation arose from the belief that eating food activated the gastrointestinal system and raised body temperature, thus negatively impacting the body if it was already suffering from a fever.

So is there any science to it? Well, yes and no. It depends on the cause of the fever and on what we are eating. Most of us aren't exactly ravenous with a high fever. That's because appetite is naturally suppressed when we are really sick (one of those sickness behaviours we met in Chapter 5) – say, when we are unable to get out of bed with a high fever. Earlier, I explained that we need energy to fuel our immunity while it's fighting a bad infection. So why then would being sick cause us to lose our appetite? It is a protective mechanism: dropping appetite when we are running a fever is designed to limit our inflammatory responses before they cause our tissues too much damage in the attempt to fight infection. Cleverly, though, as I said, it depends on the type of infection. Eating carbs is helpful against milder infections, like viruses that cause the common cold and not much fever (so, 'feed a cold'). But overdoing those carbs while suffering from a high fever, say, during a bacterial infection of the blood, throws too much fuel on the inflammatory fire, making symptoms worse (so, 'starve a fever').

What this shows is that *if* we understand what kind of infection we are dealing with, there might be simple ways to get better quicker. Now I'm not suggesting that you stuff your face the next time you have a cold or starve if you are unfortunate enough to get a fever. Much more needs to be known before we can generalise these findings, but they do highlight how profoundly the simple act of eating can impact on our immunity.

FEAST OR FAST?

Old wives' tales aside, fasting is becoming a modern-day buzzword, the latest darling of the diet industry. But in terms of science, restricting calories in the pursuit of health is definitely not new.

Eating is a metabolically active time for the body. But it's also a time of immune-system activity. When we eat, we don't just take in nutrients – we trigger the immune system to produce a transient inflammatory response. This means that just the act of eating each meal causes a degree of inflammation. Oxidants and free radicals produced during the metabolism of food can activate inflammatory gene pathways, particularly if we are eating energy-dense foods or just eating all the time, creating a metabolic traffic jam. On top of this, for around four hours after each meal, that leaky gut (see p. 98) causes gut microbes and their components to leak into the bloodstream, silently triggering inflammation by the immune system, driven largely by the activation of the inflammasome (see p. 253). This inflammation after eating – known as 'postprandial inflammation' – is normal and transient. A fibre-rich and phytonutrient-dense diet seals the barrier up again, while periods without food in between meals strengthen it. But leaky gut can be exacerbated by calorie-dense meals, frequent eating, excessive fructose and fatty foods – particularly saturated fat. This is like recurrent collateral damage to our bodies that is extremely detrimental to health over time.

Fasting Takes Many Forms

Before fasting hit the scientific mainstream, there was caloric restriction (CR), referring to reducing 60–70 per cent of your required calories each day without considering when you eat. Back in the 1930s scientists showed that rats on a calorie-reduced diet lived twice as long as other rats. In the decades since, similar results have been seen in an increasing number of animals from mice to fish to dogs. Since the start of my career, I can recall being at many scientific conferences, watching presentations on the benefits of CR for rejuvenating and aiding all manner of age-related ills such as cancer, heart disease and diabetes. In the life of a lab rat at least, restricting calories (by about 40 per cent without malnutrition or micro-nutrient deficiency) would appear not only to extend life, but to slow nearly all age-sensitive decrements of immune function and reduce the risk of a variety of diseases. Some studies suggest that CR may have health benefits for humans, but more research is needed before we understand its long-term effects. Currently, there are no data in humans on the relationship between CR and longevity.

Fasting, it seems, has become CR's heir apparent. In the 1940s researchers started fasting lab animals instead of CR. To their surprise they found the animals lived substantially longer and were less likely to get cancer. Experimentally, at least, it seems eating nothing is better for health than eating less. But is it too good to be true?

More work is needed to determine whether fasting or calorie-restriction diets have real potential for our immunity, health and longevity. But do these early lifespan results come from consuming fewer calories or from eating within a certain timeframe? From what we do know, any form of CR needs to be practised for prolonged periods to really shift any longevity levers. And are the results affected by what we eat as well as when? While science grapples with the details, the recent spotlight on fasting has seen all sorts of calorie-restricting patterns spring up with various permutations of when we eat and how much:

- Water-only fasting diet is where a person either does not eat at all, normally for a period of longer than 48 hours, or severely limits intake during certain times of the day, week or month. An additional effect of fasting may be fewer calories because there is less time for regular eating.
- Intermittent fasting (IF) focuses on when rather than what you eat. This could be done through alternate-day fasting or, 'the OMAD' (one meal a day; as the name suggests, you eat one meal every day and spend the rest of the 24 hours fasting).
- TRE (time-restricted eating) is another popular format, whereby all of your meals and snacks are eaten within a particular window of time each day. This can vary from a 12-hour window to the popular 16/8 method: 16 hours of no food and you consume all your meals within an 8-hour window. Moulding TRE around circadian biology stems from the idea that our metabolism follows a daily rhythm. If you remember from Chapter 4, eating is a zeitgeber that helps keep our biological clock on time. A growing body of research suggests that our bodies function optimally when we align eating patterns with our innate 24-hour cycles. The gut has a clock that regulates the daily ebb and flow of enzymes, the absorption of nutrients and the removal of waste. The communities of trillions of bacteria that comprise the microbiomes in our guts operate on a daily rhythm as well. What is fascinating about this is that it demonstrates that a key function of the immune system

directly depends not only on what and how much but when we eat too.

Can Going without Food Really Make You Healthier?

Throughout our evolution, finding food has been, at times, unpredictable. As a result, we have quite a few checks and balances in place to tolerate periods without food. Since spells of feasting were much rarer, we have fewer measures for too much food, which partly explains obesity becoming a health problem as our food environment has changed. So far, fasting is undoubtedly a potentially powerful *experimental* tool, but the various mechanisms at play are still being worked out. Most research to date has focused on the weight-loss aspect of fasting (it's easier to under-eat when there are fewer opportunities to eat). But the inflammation-controlling and immune-invigorating properties of fasting have more recently become sensationalised, with media reports that fasting can regenerate entire organs savaged by autoimmunity. So what's the truth behind the headlines?

Being without food is a form of stress. As you have seen, stress ushers in protective reactions in our biology. Restricting calories activates genes that direct cells to preserve resources. Rather than grow and divide, cells in famine mode are, in effect, stalled. In this state, they also upregulate proteins that help them resist disease and further stress. They are forced into a form of self-recycling known as autophagy. This process of cleaning out dead or toxic cell matter and repairing and recycling damaged components gets rid of older, worn-out immune 'zombie' cells (see p. 84), which are more likely to go wrong. When you starve, the system tries to save energy. Damaged or old cells are much more vulnerable to recycling, which is considered to be one of the mechanisms by which fasting can be helpful to our health. The neat thing is that after a fast, eating jump-starts production of fresh new immune cells from the bone marrow, replacing the damaged ones. In these experimental conditions, fasting cycles flip a switch, regenerating the immune system. Failure to do this housekeeping regularly has been implicated in autoimmune diseases and cancer. But don't worry, fasting isn't the only way our bodies take out the trash – sleep and exercise can potentially help with this too.

There are studies linking fasting to improvements in several components of immune function. Autophagy also seems to play a central role in steering the various flavours of our immunity, and its ability to detect and seek out infections. It can minimise the age-associated zombie cells that you met in Chapter 2 and reduce unwanted inflammation.[20] Reduced autoimmune T helper 17 immune cells have been shown in people eating one meal a day, which raises hope for sufferers of autoimmune conditions. Cancer-surveillance NK cells appear to work better at seeking out tumour cells. And it was recently reported that restricting calories on alternate days improved asthma symptoms in patients.

As it stands, fasting in one form or another could reset a broken immune system. This holds enormous potential to treat, even perhaps cure, autoimmune disease – and possibly in preventing and treating some cancers too. Despite this encouraging work, however, not all the research is in agreement. Autophagy has also been demonstrated to have damaging effects in autoimmune diseases. In rheumatoid arthritis, activation of autophagy upregulated pro-inflammatory signals, promoting the destruction of the joint architecture. Similarly, dysregulation of autophagy signalling has been implicated in lupus and Crohn's disease. Plus autophagy is highly complex. We don't know exactly when these processes happen and where in the body (do all cells undergo autophagy at the same time?). Nor are there any good ways to measure it yet. With most of the studies conducted in laboratory and animal models, from yeast cells to primates, the findings do not necessarily directly apply to humans. True fasting triggers survival mechanisms in small mammals with short lifespans. But for us humans, the same might not be true.

The Dark Side of Fasting

While science tries to work out these kinks, you might be enticed by the constant flow of anecdotes in support of fasting protocols. But you must guard against undernourishment, muscle loss, potential gallstones, reduced bone density, hormonal changes and crashing thyroid function, and deal with fatigue and reduced capacity for exercise, not to mention sleep loss and the psychological toll. Some studies have shown that a 48-hour total fast induced parasympathetic withdrawal with simultaneous sympathetic stress activation, reflecting a huge increase in cortisol, the stress hormone we met earlier. If you are already struggling under a heavy life load, then further increasing your stress response with a fast might just tip you over the edge. Some experimental studies show that eating a radically restricted diet may weaken the immune system and even shorten life. An interesting study in fruit flies found that those given half their normal diet and exposed to a form of salmonella lived almost twice as long as their full-fed brethren; but when infected with listeria, another food-poisoning bug, the dieting flies died after just four days, compared with six or seven for those eating normally.

This should raise a cautionary flag for those hoping to live longer by eating less. But if, like me, you are a curious self-experimenter and want to give it a try, then consuming your daily calories at regular mealtimes within a specific time window might be a manageable place to start.

DIET PATTERNS

If you're looking to rethink your eating habits for health, you might want to step back from focusing on individual foods and nutrients. Why?

Well, these days the burden of disease for nutrition has shifted; no longer are nutritional-deficiency illnesses like scurvy and rickets the prevailing diet-induced disease state. Instead, overnutrition and diets devoid of fibre and phytonutrients are the silent drivers of chronic inflammatory disease. One of the best tools we have when eating for health is looking at holistic dietary patterns, not nutrients. After all, rarely do we eat nutrients in isolation. Food is always the sum of its parts, not just a collection of macro- and micronutrients. So it's not about being inflexible or excluding food groups, but reflecting the totality of all foods and drinks consumed, with nutritional needs met primarily from foods through adaptable healthy-eating patterns. But food also has structure and texture, social and cultural implications, environmental and political undertones. Pieced together, this affects when we eat, how much we eat and how our bodies respond to it.

THE MAKING OF A SAD

The Western pattern diet (WPD), often referred to as the standard American diet (SAD), is a modern dietary pattern common in Western countries and one quickly being adopted across our increasingly globalised world. While numerous changes in lifestyle constitute modern living, SAD eating is fast gaining attention as a contributor to the increase in immune-mediated diseases. Generally, the SAD is defined as high in total energy (calories) and a high intake of saturated fats (from fatty domesticated and processed meats), refined grains, sugar, alcohol and salt, with an associated reduced consumption of fruits and vegetables and a low intake of fibre.

SAD eating is strongly patterned by socioeconomic status. This means it is part of a bigger problem that is beyond the scope of this book, though worthy of a short mention. The last few decades have forced a shift towards reliance on SAD as our lives have become more sedentary, busier and more stressful. Slowly, in the time since this cultural shift, we've seen health issues emerging, highlighting the slow incubation time of chronic lifestyle-driven disease.

This means that the calories in SADs are frequently from 'processed' foods. Now the definition of a 'processed' food is controversial. Nearly all the food at grocery stores is subject to some processing: pasteurised, vacuum sealed, cooked, frozen, fortified and mixed with preservatives and flavour enhancers. Some of these processes can change a food's nutritional qualities. To be considered 'minimally processed', foods can be frozen, dried, cooked or vacuum packed, but not include added sugar, salt or oil. Today, more than 50 per cent of food bought in UK households is considered 'ultra-processed', defined as ready-to-eat formulations with five or more additives, dyes or stabilisers, containing sodium, synthetic trans fats and artificial sweeteners to enhance flavour and extend shelf life. The UK National Diet and Nutrition Survey shows most of us are having too much fat, salt and sugar, and too few are eating five portions of fruit and veg a day.

What Makes SAD So Dangerous?

What we know is that people eating an ultra-processed SAD suffer poorer health than those eating more traditional ones. But now you can add a compromised immune function and potential higher risk for inflammatory disease. Although most inflammatory conditions have complex causes, SAD eating patterns can accelerate them and make symptoms worse.[21] Despite clear links to poor health, though, it has been challenging to pin down what exactly is so bad about the SAD.

Too Fatty, Too Sweet or Too Salty ...

The immune system reacts to typical SADs in the same way it would to a bacterial infection – with unruly inflammation. The refined, sugary carbs, saturated and trans fats, topped off with too much salt, can all independently act as danger signals, pulling the trigger on inflammasome activation, setting in motion inflammatory responses and long-term immune-system stimulation that could play a role in the development of diet-related problems such as diabetes and heart disease. When this eating pattern is persistent, the danger doesn't stop and neither does inflammation.

As you read earlier, saturated and trans fats are bad news when it comes to our immunity, as is poor blood-sugar control.

As for salt – it's vital for human functioning, and it's pretty easy to fulfil our daily needs by eating a balanced diet. The problem is that it's also easy to overconsume, and not so easy to avoid in our modern-day diets. A recent study found salt consumption too high in all but 6 of 187 countries investigated. The average adult in the UK eats about 8g a day – 2g more than the recommended upper limit.

Research has provided strong evidence that excessive salt could be involved in causing and increasing the severity of autoimmune diseases.[22,23] Smokers who consume excessive salt have double the risk of developing rheumatoid arthritis.[24] And over-salting food strips away the protective mucus from your gut, so it's quite unfriendly to your good bugs too. So it's hard to argue against being a bit more salt savvy with your diet.

The SAD may be hurting our health by being too sweet, too fatty and too salty, but the harmful effects are perhaps also due to the absence of phytonutrients, fibre and omega-3 fatty acids in SAD diet patterns, which we know to be so vital for health. Plus, our gut microbiota don't like junk food either. Food additives, specifically emulsifiers, impact them, and it may be that the increased risk for inflammatory disease seen by consuming the SAD is due to the far-reaching impacts of fibre deficiency in starving off vital gut bugs, not because of the much-debated high sugar, salt or fat content. More disturbing is the observation that harmful dietary effects can influence future generations. We know a mother's diet may alter many health factors in her unborn children, even down to flavour preferences and diet choices.

... And Too Tasty

Another big issue with the SAD is that it is just so damned tasty. The formulation of ultra-processed foods really hits the bliss point – the right amount of salt, sugar or fat to optimise deliciousness and hyper-palatability. The human body has evolved to favour foods delivering these tastes; the brain responds with a 'reward' in the form of endorphins, remembers what we did to get that reward and makes us want to do it again. The texture and taste of many processed foods means it's easier to consume them quickly too, which has a big part to play in our health. Faster eating makes it easy to consume more calories than you are using, which is a key factor in weight gain.

Caloric balance is like a scale. To remain in balance, the calories consumed (from foods) must be balanced by the calories used (in normal body functions, daily activities and exercise). SAD eating can easily lead to overconsumption of calories (known as overnutrition – a form of malnutrition):

nutrient-poor foods supplied in excess of the body's needs. The impact of overnutrition is often not felt over days or weeks, but typically accumulates over periods of years, and is damaging to immunity in many ways. Regularly eating to overload creates an imbalance between energy intake and energy expenditure, which we notice as weight piling on. Our cells don't know what to do with all the extra energy they are being fed, and then there is the excessive metabolic waste products they produce, linked to chronic inflammation. Fat cells become stressed trying to deal with too much energy. They release inflammatory substances that can act as false alarms, which, over time, can cause the immune system to dial down its responsiveness to infection, while at the same time setting the stage for the slow burn of chronic inflammation. This is one of the reasons why obesity itself is a major factor in declining immunity.

What's the Deal with Snacking?

Forty years ago, the majority of people ate at home, with most acquiring basic cookery skills to prepare healthy, balanced meals. Generally, they had a routine: three meals a day, substantial and hearty portions at socially defined times. Snacking wasn't the thing it is today. There were no school or workplace vending machines and few corner shops selling fried fast foods. Today, however, there is little time to cook, but 'luckily' food at every turn. We eat on the go with no need to cook to survive.

The changing food environment in the last few decades has been designed to meet specific objectives: to be cheap and tasty, easy to mass produce, portable, with a generous shelf life and easy to eat. I include 'healthy' packaged junk food here too. There are also fewer rules about what to eat and when. Eating is driven by availability, wants and whims, aspirations and ethics. We're more conscious of health outcomes when choosing what to eat, and while we may idealise having three balanced meals a day, we rarely eat that way.

We are also eating more frequently than ever – often outside of mealtimes. New smartphone app data confirms that we have erratic eating patterns, with half of all eating occasions defined as 'snacks'.[25] Many of us spend up to 16 hours a day in a 'fed' state. This changing eating pattern lends itself to overnutrition, jamming up metabolic traffic and entering an inherently inflammatory state. And we may have forgotten that it's OK to feel a little hungry sometimes – pathologised it, even. So it might be worth consolidating food into fewer, more satisfying meals and asking ourselves if we really need that snack.

Not all snacking is bad, of course. Satiety, the feeling of fullness that persists after eating, is an important factor in stopping us from overconsuming at mealtimes. Eating snacks between meals has the potential to promote satiety and contribute to our daily nutritional needs. Whole foods high in protein, fibre and whole grains (nuts, yoghurt, fruits, vegetables, complex carbs) improve satiety when consumed as snacks. But the majority of snack foods today are ultra-tasty, salty or sweet, energy-dense and nutrient-poor – even some of those marketed as 'healthy'. So the bodies of people snacking around the clock can end up in a near-constant inflammatory state.

DO YOU NEED TO SNACK?

I am known by friends and family as the anti-snacker. But what I really want people to do is to tune into signals:

- Are you hungry? Or is it boredom or the fact that you have food all around you that is making you want to eat?
- How's your energy? Was your last meal balanced and substantial, or did you deliberately deprive yourself through trying to be 'good'?
- Body feedback – are you experiencing tummy rumbling and a loss of concentration? Ideally, your meal should come before you have got to the shaky/low-blood-sugar feelings.

Despite the well-known public-health messages, the SAD is hard to escape due to the 24/7 drip of food. Convenience is nice, but it comes at a cost. The onus of portion control is on us. And it has never been more challenging.

THE MEDITERRANEAN DIET

The term 'Mediterranean diet' was born out of early health data from the 1960s correlating low risk of heart disease, neurodegeneration and cancer to describe an eating style inspired by the traditional dietary patterns observed in Greece and southern Italy. Typically high in fresh fruits and vegetables, whole grains and olive oil, and low in saturated fat, it has been intensely examined ever since. And with thousands of research papers, the Mediterranean diet is now perhaps the best studied and generally healthful diet pattern globally.

The evidence is that this way of eating *and* living results in an across-the-board reduction of chronic disease and offers longevity benefits. It's important to remember that there is no one format of Mediterranean eating – rather, several variations through the 18 countries that border the Mediterranean. All of these countries have eating patterns focused on high-quality food choices and limited processed foods, with the following characteristics:

- **Nutrient dense.** Phytonutrient- and fibre-focused; abundant in fruits and vegetables; breads and other forms of whole grains and cereals; and beans, nuts and seeds.
- **Minimally processed.** Originally, the Mediterranean diet was an eating pattern of the poor and, as such, it includes locally grown, seasonally fresh foods and use of beans and pulses as a cheaper way to spin out dishes and enjoy meat-free protein sources.
- **Limited sweet treats.** Fresh fruits are the typical daily dessert, with sweets based on nuts, honey and made with olive oil.
- **High-quality fats.** Olive oil is the primary fat used in cooking, with other dietary fats coming from the sea. Total fat intake is moderate.
- **Moderate dairy intake.** Mainly from cheese and yoghurt.
- **Protein.** Meats are consumed within reason and with low frequency; seafood intake varies, with moderate amounts of fish.
- **Herbs and spices.** These are used to add flavour to foods

Anti-Inflammatory Eating

Just as there is no one food or nutrient that makes the SAD pattern of eating so bad, the health benefits of a Mediterranean-style diet pattern are complex. Rather than being the result of individual foods, the benefits stem from the combination of foods, or just from eating more high-quality, nutrient-dense foods rich in phytonutrients, fibre, good fats and flavour, along with their collective antioxidant, antimicrobial and anti-inflammatory properties, while mostly steering clear of ultra-processed foods.

Much of the research on the Mediterranean diet focuses on metabolic and heart health. In people with type 2 diabetes, the majority of studies report that this pattern of eating improves blood-sugar control and chronic risk factors. In people at risk for diabetes, studies report it to be protective. Even blood-pressure medications work better for those following a form of a traditional Mediterranean diet. It's also protective against several diseases associated with chronic low-grade inflammation, such as cancer and autoimmunity, although more research is needed to determine whether such results can be achieved outside the Mediterranean geographical region.

The Mediterranean dietary pattern of eating is anti-inflammatory in nature.[26,27] It is by no means the only diet pattern intertwined with health, but it is perhaps one of our best weapons in preventing and treating the rising tide of inflammatory and non-communicable diseases.

'Gioie della Tavola' – The Joys of the Table

While the focus has been on food, eating Mediterranean style also involves a cultural aspect that we may not be able to recreate elsewhere. With half of my family having Neapolitan roots, I have often experienced the *gioie della tavola*: family bonds (and battles), dramas and celebrations, and the general joy, warmth and magic created around the Italian table. With this comes time for proper chewing and digestion that may avert digestive upset from quickly gulping down food as we increasingly do in our fast-paced lives.

Though culturally diverse in their recipes, flavour pairings and ingredients, nearly all traditional diets across the globe are centred on a foundation of fruits, vegetables, whole grains, pulses, nuts, seeds, herbs, spices, fibre and other plant-based foods. The modern-day Scottish diet is perhaps not renowned for being the healthiest. But when I consider my healthy 90-year-old grandparents in Scotland, still starting their day with porridge topped with a splash of full-fat cold milk, then lunch which is a homemade seasonal soup and bread, with a hearty dinner to end the day (always taking time to enjoy a cake and, at times, a wee dram of whisky) I can't help thinking it's less about what's being eaten and more about the general pattern: regular meals with a blueprint that focuses on phytos, fibre, good fats and flavour, extending beyond just breaking bread to a whole-life experience.

The food we eat is just part of a multi-faceted set of choices we make (or have made for us) around how we live and behave. With so many of the values that make traditional diets a healthy way of living rapidly fading, eating the Mediterranean way is only going to get us so far.

A WORD ON ALCOHOL

In many healthful dietary patterns, such as the Mediterranean, wine is consumed in low quantities with meals. In terms of immunity, regular alcohol intake may be bad for our defences.

First, immunity could indirectly be affected through alcohol's influence on sleep, which may be poorer in quality and quantity. In terms of gut health, alcohol can cause serious problems. Gut dysbiosis is more common in those who drink regularly[28] and hard spirits (like gin) particularly decrease the number of beneficial gut bacteria. Red wine, however, has been shown to increase the abundance of bacteria known to promote gut health, while decreasing the number of harmful gut bacteria.[29] The beneficial effect of moderate red wine consumption on gut bacteria appears to be due to its content of phytonutrients, in particular polyphenols.

How the body responds to alcohol depends on several factors. It is actually immune cells in the brain that are responsible for the clumsiness we associate with alcohol! Enjoying an occasional glass of organic red wine will have a completely different impact from binge drinking hard spirits all weekend. While moderate alcohol consumption may have some health benefits, higher amounts can lead to severe problems. People who drink alcohol excessively tend to be at an increased risk for infectious diseases, take longer to recover from illnesses and have more complications after surgery. Heavy alcohol intake can also affect organs that regulate immunity, such as the liver, which produces antibacterial proteins that ward off bacterial diseases and bone-marrow stem cells, which make new immune cells. This explains in part why we tend to get sick during the Christmas party season or after a heavy weekend of partying. If you do suffer from a chronic disease, alcohol may have a cumulative effect and make symptoms worse.

FOOD ALLERGY VS FOOD INTOLERANCE

It's normal to suffer some sort of digestive discomfort after eating every now and then. This might range from trivial, infrequent or frustrating to restrictive and sometimes frightening. Any reaction beyond mild discomfort occurring after eating is called an adverse food reaction. It can be the result of either a food intolerance (often referred to as sensitivity) or an actual food allergy. Today, people are quick to jump to the conclusion that they are suffering from some kind of adverse food reaction, leading them to exclude many common staples from their diets. But what's really going on? Allergy or an intolerance? And, more importantly – why does it matter?

The distinction between food allergy and food intolerance can be confusing, and it's easy to see what the misunderstanding stems from – both food allergies and intolerances involve, well ... food. There is a grey area between their many shared symptoms: bloating, nausea, stomach pain, diarrhoea, vomiting even rashes, joint pain and headaches. But this is where the similarities end. Allergic reactions to food can be life-threatening and require strict exclusion of the offending food, whereas this might not be the best approach for food intolerances in the long term, so it's important to speak with a healthcare professional before removing suspect foods from your diet.

True Food Allergies

True food allergies are rare but potentially deadly, triggered by even trace amounts of the trigger food. Best estimates suggest that less than 4 per cent of UK adults have true food allergies. Unlike food intolerances, allergies are caused by the immune system. They are the result of normal immune defences usually reserved for fighting infections going awry and inappropriately reacting to a harmless component of food.

Normally, our digestive systems interact with the food we eat without generating an immune response. This is because the immune system has switches in place that recognise that food does not pose an infectious threat, so it 'tolerates' the food and remains switched off. This is termed 'immunological oral tolerance'. Sometimes, a failure to develop oral tolerance or a breakdown in oral tolerance leads to our immune system mistakenly seeing a food as a threat and becoming sensitised to it, responding by releasing a particular type of antibody known as immunoglobulin E (IgE), which sticks to mast cells lining our digestive tracts. Here, they are poised and ready to mount an inflammatory reaction when the offending food is next eaten.

Once we have an allergy, the symptoms can run the gamut from annoying hives and swelling to a life-threatening whole-body reaction known as anaphylactic shock. Among the most commonly offending foods are peanuts, tree nuts, cow's milk, egg, soy, wheat, shellfish and fish. Oral allergy syndrome (OAS), also known as pollen–food allergy syndrome, is caused by cross-reacting allergens found in both pollen and raw fruits, vegetables or some tree nuts. If you have hay fever and feel like you're always getting an itchy mouth after eating the same raw fruit or vegetable, then it's possible this is what you have. Mostly, OAS is mild and does not cause a severe problem. Cooking foods will denature the offending proteins and is one way to avoid symptoms.

Unlike food intolerances, there are *valid* evidence-based tests for food allergies, which can include a skin-prick challenge: a small, diluted amount of the suspected food component is placed on the skin and the skin is then pricked. If a small swelling appears (in conjunction with a detailed clinical history), IgE-mediated food allergy may be diagnosed. But the current gold standard for a true food-allergy diagnosis is a specific IgE blood test in the context of a medical history, followed by a double-blind placebo-controlled food challenge. This is where suspect foods are given, but both patient and practitioner are unaware of which ones. Because of the life-threatening nature of food allergies, these tests are normally performed by registered health professionals in medical facilities. Oral antihistamines are used to treat mild reactions and injectable adrenaline or resuscitation for severe cases. Treatment of a true food allergy is usually complete exclusion, although immunotherapy is sometimes an option. This involves giving gradually increasing doses of extracts of the offending food as an injection or under the tongue. It is currently only being tested and trialled under research conditions because of the risk of anaphylaxis.

Food Intolerances

Food intolerances and some difficult-to-diagnose gut conditions are easily confused with allergies. Estimates suggest up to 20 per cent or more of people alter their diets due to a perceived adverse reaction to food.[30] Self-diagnosis is common, including by people who don't have an allergy but think they do, and those who have had an allergic reaction but blame the wrong cause. A study of 969 children found 34 per cent of parents reported food allergies in their children, but only 5 per cent were found to have one. Implementing unnecessarily restrictive diets is leading to health problems, in terms of malnutrition and more broadly.[31,32]

A food intolerance is a non-immune response of your digestive system to a food. While symptoms vary, they tend to be insidious, mostly involving the digestive system – diarrhoea, bloating, reflux, nausea and constipation – skin and respiratory system. Aches and pains, rashes, lethargy and anxiety, headaches and runny noses are typical too. Confusingly, many of these are common medically unexplained symptoms, also seen in conditions like fibromyalgia and chronic fatigue syndrome.

If you have an immediate reaction after eating a food, it is more likely to be an allergic response. With food intolerance, symptoms can occur quite quickly but are often delayed by up to 48 hours and last for hours or even days.[33] This makes the offending food especially difficult to pinpoint. What's more, if you frequently consume foods that you are intolerant to, it may be hard to attribute symptoms to a specific one. A food diary can be useful in helping you to narrow down potential intolerances.

Intolerances generally occur because your body cannot break down food properly during digestion or the food irritates your digestive tract. Food intolerances are broadly categorised as physiological (e.g. due to a true enzyme deficiency in the case of lactose), functional (such as irritable bowel syndrome in response to dietary components like fructans – see p. 262) or pharmacological (sensitivity to food additives or naturally occurring components of foods – e.g. sulfites). Intolerances can also be psychological (in the case of an eating disorder) or, in some cases, foods trigger adverse reactions without a defined cause (known as idiopathic). Microbiome disturbances and eating habits (e.g. not chewing food properly) can also contribute to digestive issues.

Rethinking Intolerance and Debunking Avoidance

While short-term removal of a suspected offending food may give immediate relief, in most cases people can eat small amounts without it causing problems. And some experts suggest that complete avoidance is not advisable.

For example, most of us lose the enzyme to digest milk by the age of five or six, but good gut bacteria feed on the lactose, digesting it for us, and if we avoid dairy, the gut's population of these helpful lactose-gobbling bacteria diminishes, exacerbating intolerance.[34] So if you are lactose-intolerant, you might never be able to down a large glass of milk, but small amounts regularly can help with tolerance. So far, this has only been examined for lactose, but the same rules may well apply to other foods like gluten. Our gut biomes adapt to what we eat, so avoiding things could lead to adaptations that make food more difficult to digest, and may open up new and unexpected side effects.

Unguided avoidance is also problematic because it could put you at risk of nutrient deficiencies. What's more, it may feel unmanageable or mask other problems or anxieties. A developing concern is that a significant proportion of people who embark on highly restrictive diets because of a perceived food intolerance may actually represent a newly described form of eating disorder:[35] avoidant/restrictive food intake disorder (ARFID) has recently been included in the *Diagnostic and Statistical Manual of Mental Disorders*,[36] reflecting recognition of the growing number of people who exhibit food-avoidant and restrictive eating behaviours.

While food intolerance is a real condition that deserves to be taken seriously, with such a broad range of potential triggers and causes, diagnosis is complicated. To make matters worse, there are many non-valid 'tests' claiming to diagnose food allergy and intolerance available commercially on the high street and on the internet – from applied kinesiology to IgG blood tests – and it is difficult to know which, if any, have any evidence of diagnostic validity.[37,38,39]

Much of 'the science' of food intolerances is driven by reported increases in IgG concentrations with certain foods. These concentrations are common in asymptomatic populations, have not been shown to be causally associated with symptoms in controlled studies and are considered clinically 'irrelevant', since you don't have to be allergic or intolerant to a food to create IgG antibodies to it.[40]

Currently, diagnosis and treatment strategies have little to offer if a true food allergy has been ruled out. 'Elimination and challenge' with a healthcare professional is the best approach. Elimination diets remove foods most commonly associated with intolerances for a period of time, normally around two weeks, or until symptoms subside. Foods are then reintroduced one at a time, while monitoring for symptoms. This takes a great deal of time and commitment on the part of both patient and practitioner, but with careful observation individuals can identify the types of foods and amounts they can tolerate without experiencing symptoms. For example, people with lactose intolerance can often target up to 7g (around half a cup of milk). A number of studies have also found that having small amounts of offending foods with meals reduces the chances of experiencing symptoms after consumption. Hydrogen/methane breath tests are useful for identifying particular intolerances, such as lactose and fructose malabsorption. They involve drinking a test substance of fructose or lactose and having regular breath tests over a two- or three-hour period. If there is a malabsorption problem, excessive hydrogen or methane will be exhaled in the breath. But not everyone will produce detectable amounts, so this test is not 100 per cent accurate.[41]

Most of us can eat whatever we want with no ill effects, providing we eat it in moderation. But we have come to pathologise fullness, mistaking it for bloating or an adverse reaction to food. If you are experiencing symptoms that you believe to be caused by a food intolerance, it's understandable that you may want to find a quick answer on Dr Google or turn to expensive yet perhaps not validated testing kits. But before deciding that you have an intolerance and embarking on a self-prescribed exclusion diet, aim to record your symptoms in a food diary and consult with a qualified health professional who will have more sophisticated tools than Google to help you.

A QUICK GUIDE TO THE FACTS

A food allergy:

- is an adverse response by your immune system
- usually comes on suddenly
- is triggered by a small amount of food
- happens every time you eat the food, regardless of how much or how frequently
- can be life-threatening

A food intolerance:

- is not mediated by your immune system
- usually comes on gradually
- often relates to the amount of food eaten
- often relates to the frequency of eating the food
- is not life-threatening

IMMUNE-BOOSTING FOODS: FACT OR FICTION?

Walk through the aisles of any health-food store and you'll find no shortage of nutritional supplements, cold remedies and fortified foods that promise to 'boost' your immune system. But the notion of the immune system as some kind of internal force field that can be boosted is rooted in misunderstood science (see p. 23). And because of the way the immune system is designed to work, we wouldn't even want it to be boosted! Nevertheless, I appreciate that the term 'immune-boosting' is an unfortunate malapropism, and that the goal is mainly to avoid succumbing to seasonal lurgies and to get swiftly back on our feet (although the potential placebo effect also complicates things – for example, we feel better because we are investing in something marketed to make us feel better). So let's have a look at some of the claims.

Echinacea to Prevent Colds and Flu

Despite a large volume of evidence that it might be helpful in treating or preventing colds (including large meta-analyses), the jury is still out on echinacea. It's complicated by the fact that there are three different species of echinacea, and numerous parts of the plant all considered to contain various active ingredients. This means there are over 800 different echinacea products with little information or consensus on the best formula to take, what dose or how long for. Plus it may interact with some medications too. *E. purpurea* is the species that studies show is most beneficial for your health, and it's also the kind that was endorsed recently by the German government as a cold preventative, because it's antiviral, antibacterial and anti-inflammatory.

Infection-fighting Turmeric

Turmeric is having a moment in the health media, but should we get sucked into the hype? Despite some solid anti-inflammatory and antioxidant properties shown to be clinically effective in treating specific illness such as forms of arthritis, the claims are often overblown by the media into a catch-all cure for all ailments. Much of the research is on curcumin, one of the active ingredients in turmeric, but there are over 300 compounds in turmeric, and curcumin-free turmeric was also clinically effective, so stick to the whole root if you choose to use it. Interestingly, raw turmeric seems to be more anti-inflammatory, whereas cooked appears better for protecting from oxidative damage. Turmeric is also a good inhibitor to viral entry into our cells. So adding this spice regularly to meals could be useful in warding off infections. However, bioavailability can be a problem. Eaten with a source of fat and a pinch of black pepper, turmeric gives remarkable improvements in digestive uptake, so make sure you add these when consuming it.

Sweating It Out with Spicy Foods

Some people swear by spicy foods for healing by 'sweating it out'. Capsaicin, the chilli pepper component that produces a burning sensation, can be effective against nasal congestion and lowering inflammation, therefore reducing symptoms. It's shown clinical utility in pain management too. There is also something to be said for enjoying a nourishing curry when feeling under the weather, and the veggies and spices are a great way to get a healthy hit of antioxidants, fibre and polyphenols.

Elderberry – The Traditional Immune Booster?

Elderberries are useful in winter, and have been used for thousands of years as both a medicine and in food, purported to decrease pain and inflammation with antiviral properties. Studies do indeed show that elderberry syrup can specifically shorten the duration and reduce symptoms of respiratory infections. In fact, when trialled against the anti-influenza drug Tamiflu®, elderberry extract came out as more effective. Components in the elderberry have also shown potential to prevent viruses from actually getting into our cells in the first place. There is a caveat, though. While current research does appear to substantiate the preventative claims against infections, the studies are small-scale and mostly funded by companies producing commercial elderberry products. While this may not negate the research and clinical trials, it is certainly a conflict of interest.

Immune Super-Trio: Lemon, Honey and Ginger

Used for generations, this trio has stood the test of time, but it does not cure a cold and actual evidence for it speeding up recovery is quite thin – with the exception of honey, which was found to be more effective as a cough suppressant for children when tested against dextromethorphan (the active ingredient in most cough medicines). In fact, the National Health Service in the UK now recommends honey first, not antibiotics, for treatment of coughs based on guidelines from the National Institute for Health and Care Excellence (NICE). Scientific evidence aside, combined in a hot drink this age-old super-trio does soothe and hydrate, and may be a viable and cheaper alternative to over-the-counter pharmaceuticals. But it's no miracle worker.

Jewish Penicillin

Chicken soup has probably been part of our diets for as long as there have been chickens and soup. It was prescribed for colds in ancient Egypt and considered a powerful remedy through the Middle Ages. The 12th-century Jewish doctor Moses Maimonides recommended it for everything from haemorrhoids to leprosy, leading to its name of 'Jewish penicillin'. Despite, until recently, a lack of scientific proof, it has been the staple recommendation for people who are feeling poorly. Although not technically a supplement, chicken soup is actually one of the most effective feel-better foods out there. This could be through a number of substances, such as carnosine, enhancing the power of immune cells, while vitamins and nutrients slow the growth of mucus-stimulating neutrophils, helping to ease mucus and inflammation in the airways. Properties from the chicken released during cooking resemble the drug acetylcysteine, commonly prescribed for respiratory ailments. It might even lower blood pressure, since collagen proteins display effects similar to ACE inhibitors, at least in animal studies. It's also a comforting and nourishing way to consume veggies, herbs and spices, while keeping you well hydrated.

Garlic – Food and Medicine?

Used for centuries as both food and medicine, garlic contains compounds that improve the germ-fighting ability of immune cells and hold potential to help prevent infection in the first place. As far back as 3000 BC the Assyrians and Sumerians used garlic to treat fevers, inflammation and injuries. Virtually every study confirms that it's good for you. It's a powerful antioxidant and antibiotic that fights off strains of staphylococcus, the bug that causes staph infections. However, many of the studies demonstrating this were of poor quality and it's not clear if you need to continually eat garlic to see the beneficial effects. A recent study exploring the impact of eating aged garlic for 90 days found that participants had significantly more T cells and NK immune cells that were much better at performing their job of defending against everyday infections. Nevertheless, the way garlic is processed can really change its effects. To optimise all the phytonutrients, use the 'hack-and-hold' technique – crushing fresh garlic and letting it stand before cooking allows enzymatic conversion of alliin to beneficial allicin, the main active ingredient.

A RECIPE FOR STRONG IMMUNITY

Modern-day celebrity diets and Instagram influencers fuel our growing obsession with how the food we eat influences our health. On the science side, there are over a million scientific publications on diet and health, and over 680,000 scientific references on inflammation, with over 30,000 peer-reviewed articles published on the relationship between diet, inflammation and health. Based on this huge body of literature – and social-media influencers aside – yes, diet is important for strong immunity and can alter your risk for major chronic non-communicable diseases by affecting inflammation. But what about if you are already suffering from a condition? While being more mindful of the food you eat is a positive step, the rise of strict diet movements as a way to 'fix' health is more polarised and confusing than ever. Along with the exponential growth of the wellness market, it is little wonder that reports of life-changing diet protocols continue to emerge in the media and popular press. But we are not just what we eat, but a mish-mash of inherited predispositions and physiological quirks.

Let's be clear: diet, though it may be helpful, is not a simple solution for what are often complex problems. I know many people reading this book will be seeking a list of immune-nourishing superfoods to add to their weekly shop, but we don't have, as yet, the knowledge to promote specific foods for specific ills, nor can we identify which diet fad will fit the needs of which condition. Perhaps such specific food-based recommendations that we can all get to grips with will come in the future, although this is not to say that diet doesn't make a difference.

Even if you don't have specific vitamin or mineral deficiencies, diet, along with other lifestyle factors, can be a prime factor in influencing a condition. The picture is emerging whereby the best way to 'eat for immunity' is to eat within a calorie balance, and ensure plenty of good fats, phytonutrients and fibre with quality carbs and diverse protein sources. Basically, avoid a SAD eating pattern (see p. 272) for the most part and stop focusing on so-called superfoods. Think holistically when you look at your plate and don't neglect the roles of other healthy lifestyle behaviours too.

I have come up with a little guide to help you make subtle shifts without feeling intimidated. Easing into these changes is the best approach for long-term success.

Good Fats, Phyto(Nutrients), Fibre, Fish and Flavour

These are my foundations for strong immunity.

- **FATS** No longer the macro to be feared, fats are crucial for our health and can play specific roles in shaping our immunity. Carefully select those that are most beneficial, like polyunsaturated and monounsaturated fats.
- **PHYTONUTRIENTS** Phytonutrients would be my superfoods if I had to have one. And considering there are so many of them, there are always going to be lots to choose from. Though some people choose to divide phytonutrients into infection-fighting and inflammation-quenching, there is much overlap.
- **FIBRE** Fibre is full of good-quality carbohydrates and is a great way to nurture immunity through your microbiome.
- **FISH** Protein is important for immunity. By looking underwater for our protein, we reduce consumption of the saturated fat present in other animal sources like red meat. Fish and seafood are complete sources of protein and also carry added benefits, like being a source of omega-3s, particularly in the case of fatty fish.

- **FLAVOUR** We are programmed to seek deliciousness. Our flavour-sensing DNA has evolved to help us survive, because it helps us avoid spoiled or harmful foods and motivates us to seek the good. Eating whole foods doesn't need to be bland. Just as you want variety in the foods you eat, you should strive for variety in the spices and herbs you use to flavour your food and bring it to life.

Reclaim the Joys of the Table

Just as the nutritional quality of our food is important, so are the emotions around eating. The dinner table has long been the gathering place for friends, family and food (remember *'gioie della tavola'* – see p. 280). Somewhere along the way, however, our culture became busier and we lost sight of our table traditions. Is your dinner table more of a decoration than a destination? Do you often eat dinner on the sofa, while scrolling through Instagram stories? If this sounds familiar, you're not paying attention to what you're eating and, as a result, not recognising when you are full. I challenge you to reclaim your dinner table!

Eat Until You Are 80 Per Cent Full

Take a leaf out of traditional Japanese culture: families concentrate on mealtimes by sitting on the floor, eating together and following an old adage, *'hara, hachi, bu'*, which translates to 'eat until you are eight parts full'. The gut and the process of digestion itself is an inflammatory stress on the body – and inflammation is the key driver of ageing. So the more you eat, the larger the inflammatory toll your body has to deal with. It usually takes the brain half an hour to recognise when you're satisfied after a meal, so stopping early helps.

Learn to Cook

In this hectic day and age of convenience where you can grab food on the go at practically any time, anywhere, why would anyone decide to spend time at home cooking from scratch? But what if cooking, in and of itself, promotes health? Not just through encouraging the consumption of more fibre, phytonutrients and healthy fats, but by making you the producer as well as the consumer. Research has shown that families who cook and eat together are healthier and less likely to overeat. The decline in home cooking closely tracks the rise in obesity and all the chronic diseases linked to lifestyle. One of the greatest gifts to your health may be investing in acquiring basic cookery skills or experimenting with new recipes and flavours.

Eat with Your Eyes

My mother is a trained cook and has worked in the catering industry all her life. She has always told me people 'eat with their eyes'. And scientifically, she might just be right. Studies show that thoughtful presentation leads to a more enjoyable meal.[42] Even with basic dishes like a simple salad or steak and chips, it meant diners found the food more palatable. Maybe this is why people are showing the world what they eat by photographing every meal and posting on Instagram (though I am sure there are some pathological aspects to that too). Either way, perhaps this is motivation to make more effort at home – particularly when feeding yourself. Just think: would you plonk dinner haphazardly on a plate to serve to your friends? I'd also recommend playful arrangements of food for small kids.

Sass Up Your Season

Think food-first meals before pill-popping vitamins and minerals. Shop small and shop local, upping your intake of seasonal produce by looking for local growers and farmers' markets. Be aware of economy and environment – when foods are in season, the price can go down. And be inventive – eating seasonally forces you to cook more and challenges your creativity: google a new recipe or come up with a dish yourself. Remember also that herbs and spices are an excellent way to increase diversity and variety in your meals; and they often pack a huge phytonutrient punch too.

Ditch Labels for Overall Patterns

Not all processed foods are bad. Nor is it helpful to label food in that way. It's what you do the majority of the time that matters most. Say no to the food prison and cultivate an attitude of gratitude.

A Final Word

The immune system governs every aspect of our physical and mental wellbeing. Deep layers of interacting networks of cells and molecules are trained from birth to develop a sensitive perception of a healthy body – one that must be learned through a reference set of key inputs tuned to a state of wellness. Immune wellness is not merely the absence of a specific disease but a particular body state.

There is a worrying trend towards generic, over-simplified health recommendations, coupled with shame and guilt from trying, and failing, to achieve perfect health. Perhaps you feel frustrated as to why you haven't been able to 'cure' your immune-mediated condition via Dr Google. I hope this book makes one thing clear; immunity is complex, and health is constantly in flow and flux. Each one of us has our own unique immune identity that is in ceaseless exchange with our microbiome, as well as responding to psychosocial and environmental inputs. This has been crucial to the survival of our species in the germy world in which we have evolved. But such an elegant defence system comes at a cost.

We can no longer ignore the fact that our fast-paced, over-consuming, unrelenting lives are slowly eroding key immunity inputs, replacing them with an aberrant set that is slowly shifting this carefully crafted system out of balance. Modern life can put immunity into a slow-burning emergency state of meta-inflammation, a proxy to ill health sometimes decades down the line. The immune system can (usually) self-correct, but sometimes it can't or becomes too unbalanced.

Equally, even if we are doing everything 'right', we don't live in an ideal world. So, how can you get the best from your immunity in the face of modern life? Rather than short-term tactics, I urge you to nail the basics. Realise that health doesn't always come in a supplement bottle or a superfood meal. We are now living lifestyles in environments diverging vastly from those in which we evolved. Despite the wonderful benefits modern life brings, the future of your health lies, in some ways, in the fusion of modern life with a return to the past. We must reunite with our 'old friends', reacquaint ourselves with traditional eating patterns, and honour our intuition to fulfil the needs of our bodies and minds.

This book isn't a promise of a quick fix or total cure. Knowledge is power and education is agency. But by equipping you with the right strategy, my hope is that this book has guided you on how to cut through the noise of wellness with better knowledge, better choices, better health, even when life works against you. Let's play the long game and build strong immunity for life.

References

Chapter 1

1. Blalock, J. E. (1984) 'The immune system as a sensory organ', *Journal of Immunology*, 132(3), pp. 1067–70. Available at: http://www.ncbi.nlm.nih.gov/pubmed/6363533 (Accessed: 30 November 2019).
2. Gruber-Bzura, B. M. (2018) 'Vitamin D and influenza – Prevention or therapy?', *International Journal of Molecular Sciences*. MDPI AG. doi: 10.3390/ijms19082419.
3. Mackowiak, P. A. *et al.* (1981) 'Effects of physiologic variations in temperature on the rate of antibiotic-induced bacterial killing', *American Journal of Clinical Pathology*, 76(1), pp. 57–62. doi: 10.1093/ajcp/76.1.57.
4. Young, P. *et al.* (2015) 'Acetaminophen for fever in critically ill patients with suspected infection', *New England Journal of Medicine*. Massachusetts Medical Society, 373(23), pp. 2215–2224. doi: 10.1056/NEJMoa1508375.
5. Irwin, R. S. (2006) 'Introduction to the diagnosis and management of cough: ACCP evidence-based clinical practice guidelines', *Chest*, pp. 25S-27S. doi: 10.1378/chest.129.1_suppl.25S.
6. Pradeu, T. and Cooper, E. L. (2012) 'The danger theory: 20 years later', *Frontiers in Immunology*, 3(SEP). doi: 10.3389/fimmu.2012.00287.
7. Matzinger, P. (1994) 'Tolerance, Danger, and the Extended Family', *Annual Review of Immunology*. Annual Reviews, 12(1), pp. 991–1045. doi: 10.1146/annurev.iy.12.040194.005015.
8. Zindel, J. and Kubes, P. (2020) 'DAMPs, PAMPs, and LAMPs in Immunity and Sterile Inflammation', *Annual Review of*

Pathology: Mechanisms of Disease, 15(1), p. 2129251537. doi: 10.1146/annurev-pathmechdis-012419-032847.
9. Chen, G. Y. and Nuñez, G. (2010) 'Sterile inflammation: Sensing and reacting to damage', *Nature Reviews Immunology*, pp. 826–837. doi: 10.1038/nri2873.
10. Brown, K. F. *et al.* (2018) 'The fraction of cancer attributable to modifiable risk factors in England, Wales, Scotland, Northern Ireland, and the United Kingdom in 2015', *British Journal of Cancer*. Nature Publishing Group, 118(8), pp. 1130–1141. doi: 10.1038/s41416-018-0029-6.
11. Hunter, P. (2012) 'The inflammation theory of disease: the growing realization that chronic inflammation is crucial in many diseases opens new avenues for treatment', *EMBO Reports*, 13(11), pp. 968–970. doi: 10.1038/embor.2012.142.
12. McDade, T. W. (2012) 'Early environments and the ecology of inflammation', *Proceedings of the National Academy of Sciences of the United States of America*, pp. 17281–17288. doi: 10.1073/pnas.1202244109.
13. Lorenzatti, A. and Servato, M. L. (2018) 'Role of anti-inflammatory interventions in coronary artery disease: Understanding the Canakinumab Anti-inflammatory Thrombosis Outcomes Study (CANTOS)', *European Cardiology Review*. Radcliffe Cardiology, 13(1), pp. 38–41. doi: 10.15420/ecr.2018.11.1.
14. Bally, M. *et al.* (2017) 'Risk of acute myocardial infarction with NSAIDs in real world use: Bayesian meta-analysis of individual patient data', *BMJ (Online)*. BMJ Publishing Group, 357. doi: 10.1136/bmj.j1909.
15. Doux, J. D. *et al.* (2005) 'Can chronic use of anti-inflammatory agents paradoxically promote chronic inflammation through compensatory host response?', *Medical Hypotheses*, 65(2), pp. 389–391. doi: 10.1016/j.mehy.2004.12.021.
16. Anderson, K. and Hamm, R. L. (2012) 'Factors that impair wound healing', *Journal of the American College of Clinical Wound Specialists*. Elsevier Inc., pp. 84–91. doi: 10.1016/j.jccw.2014.03.001.
17. Hauser, R. (2010) 'The Acceleration of Articular Cartilage Degeneration in Osteoarthritis by Nonsteroidal Anti-

inflammatory Drugs', *Journal of Prolotherapy*, 2(1), pp. 305–322

Chapter 2

1. Boehm, T. and Zufall, F. (2006) 'MHC peptides and the sensory evaluation of genotype', *Trends in Neurosciences*, pp. 100–107. doi: 10.1016/j.tins.2005.11.006.
2. Wedekind, C. *et al.* (1995) 'MHC-dependent mate preferences in humans', *Proceedings of the Royal Society B: Biological Sciences*. Royal Society, 260(1359), pp. 245–249. doi: 10.1098/rspb.1995.0087.
3. Day, S. *et al.* (2016) 'Integrating and evaluating sex and gender in health research', *Health Research Policy and Systems*. BioMed Central Ltd., 14(1). doi: 10.1186/s12961-016-0147-7.
4. Liu, K. A. and Dipietro Mager, N. A. (2016) 'Women's involvement in clinical trials: Historical perspective and future implications', *Pharmacy Practice*. Grupo de Investigacion en Atencion Farmaceutica. doi: 10.18549/PharmPract.2016.01.708.
5. Van Eijk, L. T. and Pickkers, P. (2018) 'Man flu: Less inflammation but more consequences in men than women', *BMJ (Online)*. BMJ Publishing Group. doi: 10.1136/bmj.k439.
6. Úbeda, F. and Jansen, V. A. A. (2016) 'The evolution of sex-specific virulence in infectious diseases', *Nature Communications*. Nature Publishing Group, 7. doi: 10.1038/ncomms13849.
7. Arruvito, L. *et al.* (2007) 'Expansion of CD4 + CD25 + and FOXP3 + Regulatory T Cells during the Follicular Phase of the Menstrual Cycle: Implications for Human Reproduction', *The Journal of Immunology*. The American Association of Immunologists, 178(4), pp. 2572–2578. doi: 10.4049/jimmunol.178.4.2572.
8. Ngo, S. T., Steyn, F. J. and McCombe, P. A. (2014) 'Gender differences in autoimmune disease', *Frontiers in Neuroendocrinology*. Academic Press Inc., pp. 347–369. doi: 10.1016/j.yfrne.2014.04.004.
9. Hughes, G. C. (2012) 'Progesterone and autoimmune disease', *Autoimmunity Reviews*. doi: 10.1016/j.autrev.2011.12.003.

10. Laffont, S. *et al.* (2017) 'Androgen signaling negatively controls group 2 innate lymphoid cells', *Journal of Experimental Medicine*. Rockefeller University Press, 214(6), pp. 1581–1592. doi: 10.1084/jem.20161807.
11. Lorenz, T. and van Anders, S. (2014) 'Interactions of sexual activity, gender, and depression with immunity', *Journal of Sexual Medicine*. Blackwell Publishing Ltd, 11(4), pp. 966–979. doi: 10.1111/jsm.12111.
12. Lorenz, T. K., Heiman, J. R. and Demas, G. E. (2017) 'Testosterone and immune-reproductive tradeoffs in healthy women', *Hormones and Behavior*. Academic Press Inc., 88, pp. 122–130. doi: 10.1016/j.yhbeh.2016.11.009.
13. Lorenz, T. K., Demas, G. E. and Heiman, J. R. (2017) 'Partnered sexual activity moderates menstrual cycle–related changes in inflammation markers in healthy women: an exploratory observational study', *Fertility and Sterility*. Elsevier Inc., 107(3), pp. 763–773.e3. doi: 10.1016/j.fertnstert.2016.11.010.
14. Ghosh, M., Rodriguez-Garcia, M. and Wira, C. R. (2014) 'The immune system in menopause: Pros and cons of hormone therapy', *Journal of Steroid Biochemistry and Molecular Biology*. Elsevier Ltd, pp. 171–175. doi: 10.1016/j.jsbmb.2013.09.003.
15. Rowe, J. H. *et al.* (2012) 'Pregnancy imprints regulatory memory that sustains anergy to fetal antigen', *Nature*, 490(7418), pp. 102–106. doi: 10.1038/nature11462.
16. Carr, E. J. *et al.* (2016) 'The cellular composition of the human immune system is shaped by age and cohabitation', *Nature Immunology*. Nature Publishing Group, 17(4), pp. 461–468. doi: 10.1038/ni.3371.
17. Swaminathan, S. *et al.* (2015) 'Mechanisms of clonal evolution in childhood acute lymphoblastic leukemia', *Nature Immunology*. Nature Publishing Group, 16(7), pp. 766–774. doi: 10.1038/ni.3160.
18. Ciabattini, A. *et al.* (2019) 'Role of the microbiota in the modulation of vaccine immune responses', *Frontiers in Microbiology*. Frontiers Media S.A. doi: 10.3389/fmicb.2019.01305.
19. Di Mauro, G. *et al.* (2016) 'Prevention of food and airway allergy: Consensus of the Italian Society of Preventive and

Social Paediatrics, the Italian Society of Paediatric Allergy and Immunology, and Italian Society of Pediatrics', *World Allergy Organization Journal*. BioMed Central Ltd. doi: 10.1186/s40413-016-0111-6.

20. Stein, K. (2014) 'Severely restricted diets in the absence of medical necessity: The unintended consequences', *Journal of the Academy of Nutrition and Dietetics*. Elsevier, 114(7), pp. 986–987. doi: 10.1016/j.jand.2014.03.008.
21. Chan, E. S. *et al.* (2018) 'Early introduction of foods to prevent food allergy', *Allergy, Asthma & Clinical Immunology*, 14(S2), p. 57. doi: 10.1186/s13223-018-0286-1.
22. Shaker, M. *et al.* (2018) '"To screen or not to screen": Comparing the health and economic benefits of early peanut introduction strategies in five countries', *Allergy: European Journal of Allergy and Clinical Immunology*. Blackwell Publishing Ltd, 73(8), pp. 1707–1714. doi: 10.1111/all.13446.
23. Eder, W., Ege, M. J. and Von Mutius, E. (2006) 'The asthma epidemic', *New England Journal of Medicine*, pp. 2226–2235. doi: 10.1056/NEJMra054308.
24. Baïz, N. *et al.* (2019) 'Maternal diet before and during pregnancy and risk of asthma and allergic rhinitis in children', *Allergy, Asthma & Clinical Immunology*, 15(1), p. 40. doi: 10.1186/s13223-019-0353-2.
25. Lang, J. E. *et al.* (2018) 'Being overweight or obese and the development of asthma', *Pediatrics*. American Academy of Pediatrics, 142(6). doi: 10.1542/peds.2018-2119.
26. Jones, E. J. *et al.* (2018) 'Chronic family stress and adolescent health: The moderating role of emotion regulation', *Psychosomatic Medicine*. Lippincott Williams and Wilkins, 80(8), pp. 764–773. doi: 10.1097/PSY.0000000000000624.
27. Baranska, A. *et al.* (2018) 'Unveiling skin macrophage dynamics explains both tattoo persistence and strenuous removal', *Journal of Experimental Medicine*. Rockefeller University Press, 215(4), pp. 1115–1133. doi: 10.1084/jem.20171608.
28. De Heredia, F., Gómez-Martínez, S., & Marcos, A. (2012). Obesity, inflammation and the immune system. Proceedings of the Nutrition Society, 71(2), pp. 332–338. doi:10.1017/S0029665112000092

29. Nishimura, S. *et al.* (2009) 'CD8+ effector T cells contribute to macrophage recruitment and adipose tissue inflammation in obesity', *Nature Medicine*, 15(8), pp. 914–920. doi: 10.1038/nm.1964.
30. Han, S. J. *et al.* (2017) 'White Adipose Tissue Is a Reservoir for Memory T Cells and Promotes Protective Memory Responses to Infection', *Immunity*. Cell Press, 47(6), pp. 1154–1168.e6. doi: 10.1016/j.immuni.2017.11.009.
31. Lynch, L. *et al.* (2016) 'iNKT Cells Induce FGF21 for Thermogenesis and Are Required for Maximal Weight Loss in GLP1 Therapy', *Cell Metabolism*. Cell Press, 24(3), pp. 510–519. doi: 10.1016/j.cmet.2016.08.003.
32. *World Report on Ageing and Health* (2015). Available at: www.who.int (Accessed: 1 December 2019).
33. Belsky, D. W. *et al.* (2015) 'Quantification of biological aging in young adults', *Proceedings of the National Academy of Sciences of the United States of America*. National Academy of Sciences, 112(30), pp. E4104–E4110. doi: 10.1073/pnas.1506264112.
34. Palmer, S. *et al.* (2018) 'Thymic involution and rising disease incidence with age', *Proceedings of the National Academy of Sciences of the United States of America*. National Academy of Sciences, 115(8), pp. 1883–1888. doi: 10.1073/pnas.1714478115.
35. Jurk, D. *et al.* (2014) 'Chronic inflammation induces telomere dysfunction and accelerates ageing in mice', *Nature Communications*. Nature Publishing Group, 2. doi: 10.1038/ncomms5172.
36. Baker, D. J. *et al.* (2016) 'Naturally occurring p16 Ink4a-positive cells shorten healthy lifespan', *Nature*. Nature Publishing Group, 530(7589), pp. 184–189. doi: 10.1038/nature16932.
37. De la Fuente, M. (2002) 'Effects of antioxidants on immune system ageing', *European Journal of Clinical Nutrition*, 56, pp. S5–S8. doi: 10.1038/sj.ejcn.1601476.

Chapter 3

1. Canny, G. O. and McCormick, B. A. (2008) 'Bacteria in the intestine, helpful residents or enemies from within?', *Infection and Immunity*, pp. 3360–3373. doi: 10.1128/IAI.00187-08.

2. Qin, J. *et al.* (2010) 'A human gut microbial gene catalogue established by metagenomic sequencing', *Nature*, 464(7285), pp. 59–65. doi: 10.1038/nature08821.
3. Maslowski, K. M. and MacKay, C. R. (2011) 'Diet, gut microbiota and immune responses', *Nature Immunology*, pp. 5–9. doi: 10.1038/ni0111-5.
4. Walker, R. W. *et al.* (2017) 'The prenatal gut microbiome: are we colonized with bacteria *in utero*?', *Pediatric Obesity*, 12, pp. 3–17. doi: 10.1111/ijpo.12217.
5. Lathrop, S. K. *et al.* (2011) 'Peripheral education of the immune system by colonic commensal microbiota', *Nature*, 478(7368), pp. 250–254. doi: 10.1038/nature10434.
6. Martín-Orozco, E., Norte-Muñoz, M. and Martínez-García, J. (2017) 'Regulatory T cells in allergy and asthma', *Frontiers in Pediatrics*. Frontiers Media S.A. doi: 10.3389/fped.2017.00117.
7. Abdel-Gadir, A. *et al.* (2019) 'Microbiota therapy acts via a regulatory T cell MyD88/RORγt pathway to suppress food allergy', *Nature Medicine*. Nature Publishing Group, 25(7), pp. 1164–1174. doi: 10.1038/s41591-019-0461-z.
8. Dominguez-Bello, M. G. *et al.* (2010) 'Delivery mode shapes the acquisition and structure of the initial microbiota across multiple body habitats in newborns', *Proceedings of the National Academy of Sciences of the United States of America*, 107(26), pp. 11971–11975. doi: 10.1073/pnas.1002601107.
9. Salminen, S. *et al.* (2004) 'Influence of mode of delivery on gut microbiota composition in seven year old children', *Gut*. BMJ Publishing Group, pp. 1388–1389. doi: 10.1136/gut.2004.041640.
10. Sevelsted, A. *et al.* (2015) 'Cesarean section and chronic immune disorders', *Pediatrics*. American Academy of Pediatrics, 135(1), pp. e92–e98. doi: 10.1542/peds.2014-0596.
11. Pannaraj, P. S. *et al.* (2017) 'Association between breast milk bacterial communities and establishment and development of the infant gut microbiome', *JAMA Pediatrics*. American Medical Association, 171(7), pp. 647–654. doi: 10.1001/jamapediatrics.2017.0378.

12. Bode, L. (2012) 'Human milk oligosaccharides: Every baby needs a sugar mama', *Glycobiology*, pp. 1147–1162. doi: 10.1093/glycob/cws074.
13. Reynolds, A. *et al.* (2019) 'Carbohydrate quality and human health: a series of systematic reviews and meta-analyses', *The Lancet*. Lancet Publishing Group, 393(10170), pp. 434–445. doi: 10.1016/S0140-6736(18)31809-9.
14. Kunzmann, A. T. *et al.* (2015) 'Dietary fiber intake and risk of colorectal cancer and incident and recurrent adenoma in the Prostate, Lung, Colorectal, and Ovarian Cancer Screening Trial', *American Journal of Clinical Nutrition*. American Society for Nutrition, 102(4), pp. 881–890. doi: 10.3945/ajcn.115.113282.
15. Dai, Z. *et al.* (2017) 'Dietary intake of fibre and risk of knee osteoarthritis in two US prospective cohorts', *Annals of the Rheumatic Diseases*. BMJ Publishing Group, 76(8), pp. 1411–1419. doi: 10.1136/annrheumdis-2016-210810.
16. McDonald, D. *et al.* (2018) 'American Gut: an Open Platform for Citizen Science Microbiome Research', *mSystems*. American Society for Microbiology, 3(3). doi: 10.1128/msystems.00031-18.
17. Sonnenburg, E. D. *et al.* (2016) 'Diet-induced extinctions in the gut microbiota compound over generations', *Nature*. Nature Publishing Group, 529(7585), pp. 212–215. doi: 10.1038/nature16504.
18. Helander, H. F. and Fändriks, L. (2014) 'Surface area of the digestive tract-revisited', *Scandinavian Journal of Gastroenterology*. Informa Healthcare, 49(6), pp. 681–689. doi: 10.3109/00365521.2014.898326.
19. Roomruangwong, C. *et al.* (2019) 'The menstrual cycle may not be limited to the endometrium but also may impact gut permeability', *Acta Neuropsychiatrica*. Cambridge University Press. doi: 10.1017/neu.2019.30.
20. Lozupone, C. A. *et al.* (2012) 'Diversity, stability and resilience of the human gut microbiota', *Nature*, pp. 220–230. doi: 10.1038/nature11550.
21. Bischoff, S. C. (2011) '"Gut health": A new objective in medicine?', *BMC Medicine*, 9. doi: 10.1186/1741-7015-9-24.

22. Dethlefsen, L. *et al.* (2008) 'The Pervasive Effects of an Antibiotic on the Human Gut Microbiota, as Revealed by Deep 16S rRNA Sequencing', *PLoS Biology*. Edited by J. A. Eisen, 6(11), p. e280. doi: 10.1371/journal.pbio.0060280.
23. Cox, L. M. *et al.* (2014) 'Altering the intestinal microbiota during a critical developmental window has lasting metabolic consequences', *Cell*, 158(4), pp. 705–721. doi: 10.1016/j.cell.2014.05.052.
24. Cox, L. M. and Blaser, M. J. (2015) 'Antibiotics in early life and obesity', *Nature Reviews Endocrinology*. Nature Publishing Group, pp. 182–190. doi: 10.1038/nrendo.2014.210.
25. Turnbaugh, P. J. *et al.* (2006) 'An obesity-associated gut microbiome with increased capacity for energy harvest', *Nature*, 444(7122), pp. 1027–1031. doi: 10.1038/nature05414.
26. Escobar, J. S. *et al.* (2017) 'Metformin Is Associated with Higher Relative Abundance of Mucin-Degrading Akkermansia muciniphila and Several Short-Chain Fatty Acid-Producing Microbiota in the Gut', *Diabetes Care*, 40. doi: 10.2337/dc16-1324.
27. Jackson, M. A. *et al.* (2016) 'Proton pump inhibitors alter the composition of the gut microbiota', *Gut*. BMJ Publishing Group, 65(5), pp. 749–756. doi: 10.1136/gutjnl-2015-310861.
28. Rogers, M. A. M. and Aronoff, D. M. (2016) 'The influence of non-steroidal anti-inflammatory drugs on the gut microbiome', *Clinical Microbiology and Infection*. Elsevier B.V., 22(2), pp. 178.e1–178.e9. doi: 10.1016/j.cmi.2015.10.003.
29. Maier, L. *et al.* (2018) 'Extensive impact of non-antibiotic drugs on human gut bacteria', *Nature*. Nature Publishing Group, 555(7698), pp. 623–628. doi: 10.1038/nature25979.
30. Tropini, C. *et al.* (2018) 'Transient Osmotic Perturbation Causes Long-Term Alteration to the Gut Microbiota', *Cell*. Cell Press, 173(7), pp. 1742–1754.e17. doi: 10.1016/j.cell.2018.05.008.
31. Hesselmar, B., Hicke-Roberts, A. and Wennergren, G. (2015) 'Allergy in children in hand versus machine dishwashing', *Pediatrics*. American Academy of Pediatrics, 135(3), pp. e590–e597. doi: 10.1542/peds.2014-2968.
32. Frenkel, E. S. and Ribbeck, K. (2015) 'Salivary mucins protect surfaces from colonization by cariogenic bacteria', *Applied and*

Environmental Microbiology. American Society for Microbiology, 81(1), pp. 332–338. doi: 10.1128/AEM.02573-14.

33. Lynch, S. J., Sears, M. R. and Hancox, R. J. (2016) 'Thumb-sucking, nail-biting, and atopic sensitization, asthma, and hay fever', *Pediatrics*. American Academy of Pediatrics, 138(2). doi: 10.1542/peds.2016-0443.
34. Mills, J. G. *et al.* (2019) 'Relating urban biodiversity to human health with the "Holobiont" concept', *Frontiers in Microbiology*. Frontiers Media S.A. doi: 10.3389/fmicb.2019.00550.
35. Brodie, E. L. *et al.* (2007) 'Urban aerosols harbor diverse and dynamic bacterial populations', *Proceedings of the National Academy of Sciences of the United States of America*, 104(1), pp. 299–304. doi: 10.1073/pnas.0608255104.
36. Meadow, J. F. *et al.* (2014) 'Indoor airborne bacterial communities are influenced by ventilation, occupancy, and outdoor air source', *Indoor Air*, 24(1), pp. 41–48. doi: 10.1111/ina.12047.
37. Kembel, S. W. *et al.* (2012) ''Architectural design influences the diversity and structure of the built environment microbiome', *ISME Journal*, 6(8), pp. 1469–1479. doi: 10.1038/ismej.2011.211.
38. Olszak, T. *et al.* (2012) 'Microbial exposure during early life has persistent effects on natural killer T cell function', *Science*. American Association for the Advancement of Science, 336 (6080), pp. 489–493. doi: 10.1126/science.1219328.
39. Riedler, J. *et al.* (2001) 'Exposure to farming in early life and development of asthma and allergy: A cross-sectional survey', *Lancet*. Lancet Publishing Group, 358(9288), pp. 1129–1133. doi: 10.1016/S0140-6736(01)06252-3.
40. Stanford, J. L. *et al.* (2001) 'Does immunotherapy with heat-killed Mycobacterium vaccae offer hope for the treatment of multi-drug-resistant pulmonary tuberculosis?', *Respiratory Medicine*. W.B. Saunders Ltd, 95(6), pp. 444–447. doi: 10.1053/rmed.2001.1065.
41. Skinner, M. A. *et al.* (2001) 'The ability of heat-killed Mycobacterium vaccae to stimulate a cytotoxic T-cell response to an unrelated protein is associated with a 65 kilodalton heat-shock protein', *Immunology*, 102(2), pp. 225–233. doi: 10.1046/j.1365-2567.2001.01174.x.

42. O'Brien, M. E. R. *et al.* (2004) 'SRL 172 (killed Mycobacterium vaccae) in addition to standard chemotherapy improves quality of life without affecting survival, in patients with advanced non-small-cell lung cancer: Phase III results', *Annals of Oncology*, 15(6), pp. 906–914. doi: 10.1093/annonc/mdh220.
43. Wiley, A. S. and Katz, S. H. (1998) 'Geophagy in pregnancy: A test of a hypothesis', *Current Anthropology*, 39(4), pp. 532–545. doi: 10.1086/204769.
44. Krishnamani, R. and Mahaney, W. C. (2000) 'Geophagy among primates: Adaptive significance and ecological consequences', *Animal Behaviour*. Academic Press, pp. 899–915. doi: 10.1006/anbe.1999.1376.
45. Hickson, M. *et al.* (2007) 'Use of probiotic Lactobacillus preparation to prevent diarrhoea associated with antibiotics: Randomised double blind placebo controlled trial', *British Medical Journal*, 335(7610), pp. 80–83. doi: 10.1136/bmj.39231.599815.55.
46. Spaiser, S. J. *et al.* (2015) 'Lactobacillus gasseri KS-13, Bifidobacterium bifidum G9-1, and Bifidobacterium longum MM-2 Ingestion Induces a Less Inflammatory Cytokine Profile and a Potentially Beneficial Shift in Gut Microbiota in Older Adults: A Randomized, Double-Blind, Placebo-Controlled, Crossover Study', *Journal of the American College of Nutrition*. Routledge, 34(6), pp. 459–469. doi: 10.1080/07315724.2014.983249.
47. Kumar, M. *et al.* (2016) 'Human gut microbiota and healthy aging: Recent developments and future prospective', *Nutrition and Healthy Aging*. IOS Press, 4(1), pp. 3–16. doi: 10.3233/nha-150002.
48. Ouwehand, A. C. *et al.* (2008) '*Bifidobacterium* microbiota and parameters of immune function in elderly subjects', *FEMS Immunology & Medical Microbiology*, 53(1), pp. 18–25. doi: 10.1111/j.1574-695X.2008.00392.x.
49. Hao, Q., Dong, B. R. and Wu, T. (2015) 'Probiotics for preventing acute upper respiratory tract infections', *Cochrane Database of Systematic Reviews*. John Wiley and Sons Ltd. doi: 10.1002/14651858.CD006895.pub3.

50. Wassermann, B., Müller, H. and Berg, G. (2019) 'An Apple a Day: Which Bacteria Do We Eat With Organic and Conventional Apples?', *Frontiers in Microbiology*, 10. doi: 10.3389/fmicb.2019.01629.
51. Dimidi, E. *et al.* (2019) 'Fermented Foods: Definitions and Characteristics, Impact on the Gut Microbiota and Effects on Gastrointestinal Health and Disease', *Nutrients*, 11(8), p. 1806. doi: 10.3390/nu11081806.

Chapter 4

1. Gallicchio, L. and Kalesan, B. (2009) 'Sleep duration and mortality: a systematic review and meta-analysis', *Journal of Sleep Research*, 18(2), pp. 148–158. doi: 10.1111/j.1365-2869.2008.00732.x.
2. Besedovsky, L., Lange, T. and Haack, M. (2019) 'The sleep-immune crosstalk in health and disease', *Physiological Reviews*. American Physiological Society, 99(3), pp. 1325–1380. doi: 10.1152/physrev.00010.2018.
3. Savard, J. *et al.* (no date) 'Chronic insomnia and immune functioning', *Psychosomatic Medicine*, 65(2), pp. 211–21. doi: 10.1097/01.psy.0000033126.22740.f3.
4. Cohen, S. *et al.* (2009) 'Sleep habits and susceptibility to the common cold', *Archives of Internal Medicine*, 169(1), pp. 62–67. doi: 10.1001/archinternmed.2008.505.
5. Westermann, J. *et al.* (2015) 'System Consolidation During Sleep – A Common Principle Underlying Psychological and Immunological Memory Formation', *Trends in Neurosciences*. Elsevier Ltd, pp. 585–597. doi: 10.1016/j.tins.2015.07.007.
6. Irwin, M. R. *et al.* (2008) 'Sleep Loss Activates Cellular Inflammatory Signaling', *Biological Psychiatry*, 64(6), pp. 538–540. doi: 10.1016/j.biopsych.2008.05.004.
7. Irwin, M. *et al.* (2003) 'Nocturnal catecholamines and immune function in insomniacs, depressed patients, and control subjects', *Brain, Behavior, and Immunity*. Academic Press Inc., 17(5), pp. 365–372. doi: 10.1016/S0889-1591(03)00031-X.
8. Vgontzas, A. N. *et al.* (2004) 'Adverse Effects of Modest Sleep Restriction on Sleepiness, Performance, and Inflammatory Cytokines', *The Journal of Clinical*

Endocrinology & Metabolism, 89(5), pp. 2119–2126. doi: 10.1210/jc.2003-031562.

9. Lentz, M. J. *et al.* (1999) 'Effects of selective slow wave sleep disruption on musculoskeletal pain and fatigue in middle aged women', *Journal of Rheumatology*, 26(7), pp. 1586–1592.
10. Ben Simon, E. and Walker, M. P. (2018) 'Sleep loss causes social withdrawal and loneliness', *Nature Communications*. Nature Publishing Group, 9(1). doi: 10.1038/s41467-018-05377-0.
11. Smith, R. P. *et al.* (2019) 'Gut microbiome diversity is associated with sleep physiology in humans', *PLOS ONE*. Edited by P. Aich, 14(10), p. e0222394. doi: 10.1371/journal.pone.0222394.
12. Smith, R. P. *et al.* (2019) 'Gut microbiome diversity is associated with sleep physiology in humans', *PLOS ONE*. Edited by P. Aich, 14(10), p. e0222394. doi: 10.1371/journal.pone.0222394.
13. Hoyle, N. P. *et al.* (2017) 'Circadian actin dynamics drive rhythmic fibroblast mobilization during wound healing', *Science Translational Medicine*. American Association for the Advancement of Science, 9(415). doi: 10.1126/scitranslmed.aal2774.
14. Pietroiusti, A. *et al.* (2010) 'Incidence of metabolic syndrome among night-shift healthcare workers', *Occupational and Environmental Medicine*, 67(1), pp. 54–57. doi: 10.1136/oem.2009.046797.
15. Stevens, R. G. *et al.* (2014) 'Breast cancer and circadian disruption from electric lighting in the modern world', *CA: A Cancer Journal for Clinicians*, 64(3), pp. 207–218. doi: 10.3322/caac.21218.
16. Vacchio, M. S., Lee, J. Y. and Ashwell, J. D. (1999) 'Thymus-derived glucocorticoids set the thresholds for thymocyte selection by inhibiting TCR-mediated thymocyte activation.', *Journal of Immunology*, 163(3), pp. 1327–33. Available at: http://www.ncbi.nlm.nih.gov/pubmed/10415031 (Accessed: 1 December 2019).
17. Cinzano, P., Falchi, F. and Elvidge, C. D. (2001) 'The first World Atlas of the artificial night sky brightness', *Monthly Notices of the Royal Astronomical Society*, 328(3), pp. 689–707. doi: 10.1046/j.1365-8711.2001.04882.x.

18. Hale, L. and Guan, S. (2015) 'Screen time and sleep among school-aged children and adolescents: A systematic literature review', *Sleep Medicine Reviews*. W.B. Saunders Ltd, pp. 50–58. doi: 10.1016/j.smrv.2014.07.007.
19. Scott, H., Biello, S. M. and Woods, H. C. (2019) 'Social media use and adolescent sleep patterns: cross-sectional findings from the UK millennium cohort study', *BMJ Open*. NLM (Medline), 9(9), p. e031161. doi: 10.1136/bmjopen-2019-031161.
20. Crowley, S. J. *et al.* (2018) 'An update on adolescent sleep: New evidence informing the perfect storm model', *Journal of Adolescence*. Academic Press, pp. 55–65. doi: 10.1016/j.adolescence.2018.06.001.
21. O'Hagan, J. B., Khazova, M. and Price, L. L. A. (2016) 'Low-energy light bulbs, computers, tablets and the blue light hazard', *Eye (Basingstoke)*. Nature Publishing Group, 30(2), pp. 230–233. doi: 10.1038/eye.2015.261.
22. CIE Technical Committee 6-15 and International Commission on Illumination (no date) *A Computerized Approach to Transmission and Absorption Characteristics of the Human Eye*.
23. Cissé, Y. M., Russart, K. L. G. and Nelson, R. J. (2017) 'Parental Exposure to Dim Light at Night Prior to Mating Alters Offspring Adaptive Immunity', *Nature Publishing Group*. doi: 10.1038/srep45497.
24. Meijden, W. P. van der *et al.* (2019) 'Restoring the sleep disruption by blue light emitting screen use in adolescents: a randomized controlled trial', *Endocrine Abstracts*. Bioscientifica. doi: 10.1530/endoabs.63.p652.
25. Kimberly, B. and James R., P. (2009) 'Amber lenses to block blue light and improve sleep: A randomized trial', *Chronobiology International*, 26(8), pp. 1602–1612. doi: 10.3109/07420520903523719.
26. Van Der Lely, S. *et al.* (2015) 'Blue blocker glasses as a countermeasure for alerting effects of evening light-emitting diode screen exposure in male teenagers', *Journal of Adolescent Health*. Elsevier USA, 56(1), pp. 113–119. doi: 10.1016/j.jadohealth.2014.08.002.

27. Rångtell, F. H. *et al.* (2016) 'Two hours of evening reading on a self-luminous tablet vs. reading a physical book does not alter sleep after daytime bright light exposure', *Sleep Medicine*. Elsevier B.V., 23, pp. 111–118. doi: 10.1016/j.sleep.2016.06.016.
28. Gominak, S. C. and Stumpf, W. E. (2012) 'The world epidemic of sleep disorders is linked to vitamin D deficiency', *Medical Hypotheses*, 79(2), pp. 132–135. doi: 10.1016/j.mehy.2012.03.031.
29. McCarty, D. E. *et al.* (2014) 'The link between vitamin D metabolism and sleep medicine', *Sleep Medicine Reviews*. W.B. Saunders Ltd, pp. 311–319. doi: 10.1016/j.smrv.2013.07.001.
30. Becquet, D. *et al.* (1993) 'Glutamate, GABA, glycine and taurine modulate serotonin synthesis and release in rostral and caudal rhombencephalic raphe cells in primary cultures', *Neurochemistry International*, 23(3), pp. 269–283. doi: 10.1016/0197-0186(93)90118-O.
31. Yamadera, W. *et al.* (2007) 'Glycine ingestion improves subjective sleep quality in human volunteers, correlating with polysomnographic changes', *Sleep and Biological Rhythms*, 5(2), pp. 126–131. doi: 10.1111/j.1479-8425.2007.00262.x.
32. Inagawa, K. *et al.* (2006) 'Subjective effects of glycine ingestion before bedtime on sleep quality', *Sleep and Biological Rhythms*, 4(1), pp. 75–77. doi: 10.1111/j.1479-8425.2006.00193.x.
33. Liguori, I. *et al.* (2018) 'Oxidative stress, aging, and diseases', *Clinical Interventions in Aging*. Dove Medical Press Ltd., pp. 757–772. doi: 10.2147/CIA.S158513.
34. Forrest, K. Y. Z. and Stuhldreher, W. L. (2011) 'Prevalence and correlates of vitamin D deficiency in US adults', *Nutrition Research*, 31(1), pp. 48–54. doi: 10.1016/j.nutres.2010.12.001.
35. Nair, R. and Maseeh, A. (2012) 'Vitamin D: The sunshine vitamin', *Journal of Pharmacology and Pharmacotherapeutics*, pp. 118–126. doi: 10.4103/0976-500X.95506.
36. Garland, C. F. *et al.* (2014) 'Meta-analysis of all-cause mortality according to serum 25-hydroxyvitamin D', *American Journal of Public Health*. American Public Health Association Inc. doi: 10.2105/AJPH.2014.302034.

37. Phan, T. X. *et al.* (2016) 'Intrinsic photosensitivity enhances motility of T lymphocytes', *Scientific Reports*. Nature Publishing Group, 6. doi: 10.1038/srep39479.
38. Leavy, O. (2010) 'Immune-boosting sunshine', *Nature Reviews Immunology*, p. 220. doi: 10.1038/nri2759.
39. Yu, C. *et al.* (2017) 'Nitric oxide induces human CLA+CD25+Foxp3+ regulatory T cells with skin-homing potential', *Journal of Allergy and Clinical Immunology*. Mosby Inc., 140(5), pp. 1441–1444.e6. doi: 10.1016/j.jaci.2017.05.023.
40. Sloan, C., Moore, M. L. and Hartert, T. (2011) 'Impact of Pollution, Climate, and Sociodemographic Factors on Spatiotemporal Dynamics of Seasonal Respiratory Viruses', *Clinical and Translational Science*, 4(1), pp. 48–54. doi: 10.1111/j.1752-8062.2010.00257.x.
41. Leekha, S., Diekema, D. J. and Perencevich, E. N. (2012) 'Seasonality of staphylococcal infections', *Clinical Microbiology and Infection*. Blackwell Publishing Ltd, pp. 927–933. doi: 10.1111/j.1469-0691.2012.03955.x.
42. Dopico, X. C. *et al.* (2015) 'Widespread seasonal gene expression reveals annual differences in human immunity and physiology', *Nature Communications*. Nature Publishing Group, 6. doi: 10.1038/ncomms8000.
43. Goldinger, A. *et al.* (2015) 'Seasonal effects on gene expression', *PLoS ONE*. Public Library of Science, 10(5). doi: 10.1371/journal.pone.0126995.

Chapter 5

1. Dantzer, R. (2009) 'Cytokine, Sickness Behavior, and Depression', *Immunology and Allergy Clinics of North America*, pp. 247–264. doi: 10.1016/j.iac.2009.02.002.
2. Iwashyna, T. J. *et al.* (2010) 'Long-term cognitive impairment and functional disability among survivors of severe sepsis', *JAMA – Journal of the American Medical Association*. American Medical Association, 304(16), pp. 1787–1794. doi: 10.1001/jama.2010.1553.
3. Harrison, N. A. *et al.* (2009) 'Neural Origins of Human Sickness in Interoceptive Responses to Inflammation', *Biological*

Psychiatry, 66(5), pp. 415–422. doi: 10.1016/j.biopsych.2009.03.007.

4. Freedland, K. E. *et al.* (1992) 'Major depression in coronary artery disease patients with vs. without a prior history of depression', *Psychosomatic Medicine*, 54(4), pp. 416–421. doi: 10.1097/00006842-199207000-00004.
5. Panagiotakos, D. B. *et al.* (2004) 'Inflammation, coagulation, and depressive symptomatology in cardiovascular disease-free people; the ATTICA study', *European Heart Journal*, 25(6), pp. 492–499. doi: 10.1016/j.ehj.2004.01.018.
6. Saavedra, K. *et al.* (2016) 'Epigenetic modifications of major depressive disorder', *International Journal of Molecular Sciences*. MDPI AG. doi: 10.3390/ijms17081279.
7. Wray, N. R. *et al.* (2018) 'Genome-wide association analyses identify 44 risk variants and refine the genetic architecture of major depression', *Nature Genetics*. Nature Publishing Group, 50(5), pp. 668–681. doi: 10.1038/s41588-018-0090-3.
8. Raison, C. L. and Miller, A. H. (2013) 'The evolutionary significance of depression in Pathogen Host Defense (PATHOS-D)', *Molecular Psychiatry*. Nature Publishing Group, 18(1), pp. 15–37. doi: 10.1038/mp.2012.2.
9. Penn, E. and Tracy, D. K. (2012) 'The drugs don't work? antidepressants and the current and future pharmacological management of depression', *Therapeutic Advances in Psychopharmacology*, 2(5), pp. 179–188. doi: 10.1177/204512531244546
10. Kessler, R. C. and Bromet, E. J. (2013) 'The Epidemiology of Depression Across Cultures', *Annual Review of Public Health*. Annual Reviews, 34(1), pp. 119–138. doi: 10.1146/annurev-publhealth-031912-114409.
11. Kohler, O. *et al.* (2016) 'Inflammation in Depression and the Potential for Anti-Inflammatory Treatment', *Current Neuropharmacology*. Bentham Science Publishers Ltd., 14(7), pp. 732–742. doi: 10.2174/1570159x14666151208113700.
12. Kappelmann, N. *et al.* (2018) 'Antidepressant activity of anti-cytokine treatment: A systematic review and meta-analysis of clinical trials of chronic inflammatory conditions', *Molecular*

Psychiatry. Nature Publishing Group, 23(2), pp. 335–343. doi: 10.1038/mp.2016.167.

13. Al-Harbi, K. S. (2012) 'Treatment-resistant depression: Therapeutic trends, challenges, and future directions', *Patient Preference and Adherence*, pp. 369–388. doi: 10.2147/PPA. S29716.
14. O'Connell, P. J. *et al.* (2006) 'A novel form of immune signaling revealed by transmission of the inflammatory mediator serotonin between dendritic cells and T cells', *Blood*, 107(3), pp. 1010–1017. doi: 10.1182/blood-2005-07-2903.
15. Halaris, A. *et al.* (2015) 'Does escitalopram reduce neurotoxicity in major depression?', *Journal of Psychiatric Research*. Elsevier Ltd, 66–67, pp. 118–126. doi: 10.1016/j.jpsychires.2015.04.026.
16. Graham-Engeland, J. E. *et al.* (2018) 'Negative and positive affect as predictors of inflammation: Timing matters', *Brain, Behavior, and Immunity*. Academic Press Inc., 74, pp. 222–230. doi: 10.1016/j.bbi.2018.09.011.
17. Barlow, M. A. *et al.* (2019) 'Is Anger, but Not Sadness, Associated with Chronic Inflammation and Illness in Older Adulthood? The Discrete Emotion Theory of Affective Aging Psychology and Aging', *Association*, 34(3), pp. 330–340. doi: 10.1037/pag0000348.
18. Cole, S. W. *et al.* (2007) 'Social regulation of gene expression in human leukocytes', *Genome Biology*, 8(9), p. R189. doi: 10.1186/gb-2007-8-9-r189.
19. Tomova, L. *et al.* (2017) 'Increased neural responses to empathy for pain might explain how acute stress increases prosociality', *Social Cognitive and Affective Neuroscience*, 12(3), pp. 401–408. doi: 10.1093/scan/nsw146.
20. Patterson, A. M. *et al.* (2014) 'Perceived stress predicts allergy flares', *Annals of Allergy, Asthma and Immunology*. American College of Allergy, Asthma and Immunology, 112(4), pp. 317–321. doi: 10.1016/j.anai.2013.07.013.
21. Wainwright, N. W. J. *et al.* (2007) 'Psychosocial factors and incident asthma hospital admissions in the EPIC-Norfolk cohort study', *Allergy*, 62(5), pp. 554–560. doi: 10.1111/j.1398-9995.2007.01316.x.

22. Calam, R. *et al.* (2005) 'Behavior Problems Antecede the Development of Wheeze in Childhood', *American Journal of Respiratory and Critical Care Medicine*, 171(4), pp. 323–327. doi: 10.1164/rccm.200406-791OC.
23. Stevenson, J. and ETAC Study Group (no date) 'Relationship between behavior and asthma in children with atopic dermatitis.', *Psychosomatic Medicine*, 65(6), pp. 971–5. doi: 10.1097/01.psy.0000097343.76844.90.
24. Dave, N. D. *et al.* (2011) 'Stress and Allergic Diseases', *Immunology and Allergy Clinics of North America*, pp. 55–68. doi: 10.1016/j.iac.2010.09.009.
25. Martino, M. *et al.* (2012) 'Immunomodulation Mechanism of Antidepressants: Interactions between Serotonin/Norepinephrine Balance and Th1/Th2 Balance', *Current Neuropharmacology*. Bentham Science Publishers Ltd., 10(2), pp. 97–123. doi: 10.2174/157015912800604542.
26. Koch-Henriksen, N. *et al.* (2018) 'Incidence of MS has increased markedly over six decades in Denmark particularly with late onset and in women', *Neurology*. Lippincott Williams and Wilkins, 90(22), pp. e1954–e1963. doi: 10.1212/WNL.0000000000005612.
27. Banuelos, J. and Lu, N. Z. (2016) 'A gradient of glucocorticoid sensitivity among helper T cell cytokines', *Cytokine and Growth Factor Reviews*. Elsevier Ltd, pp. 27–35. doi: 10.1016/j.cytogfr.2016.05.002.
28. Roberts, A. L. *et al.* (2017) 'Association of Trauma and Posttraumatic Stress Disorder with Incident Systemic Lupus Erythematosus in a Longitudinal Cohort of Women', *Arthritis & Rheumatology*, 69(11), pp. 2162–2169. doi: 10.1002/art.40222.
29. Oral, R. *et al.* (2016) 'Adverse childhood experiences and trauma informed care: The future of health care', *Pediatric Research*. Nature Publishing Group, pp. 227–233. doi: 10.1038/pr.2015.197.
30. Weder, N. *et al.* (2014) 'Child abuse, depression, and methylation in genes involved with stress, neural plasticity, and brain circuitry', *Journal of the American Academy of Child and Adolescent Psychiatry*. Elsevier Inc., 53(4). doi: 10.1016/j.jaac.2013.12.025.

31. Song, H. *et al.* (2018) 'Association of stress-related disorders with subsequent autoimmune disease', *JAMA – Journal of the American Medical Association*. American Medical Association, 319(23), pp. 2388–2400. doi: 10.1001/jama.2018.7028.
32. Zaneveld, J. R., McMinds, R. and Thurber, R. V. (2017) 'Stress and stability: Applying the Anna Karenina principle to animal microbiomes', *Nature Microbiology*. Nature Publishing Group. doi: 10.1038/nmicrobiol.2017.121.
33. 'WHO | Burn-out an "occupational phenomenon": International Classification of Diseases' (2019) *WHO*. World Health Organization.
34. Chan, J. S. Y. *et al.* (2019) 'Special Issue–Therapeutic Benefits of Physical Activity for Mood: A Systematic Review on the Effects of Exercise Intensity, Duration, and Modality', *Journal of Psychology: Interdisciplinary and Applied*. Routledge, pp. 102–125. doi: 10.1080/00223980.2018.1470487.
35. Zila, I. *et al.* (2017) 'Vagal-immune interactions involved in cholinergic anti-inflammatory pathway', *Physiological Research*. Czech Academy of Sciences, pp. S139–S145.
36. Morrison, I. (2016) 'Keep Calm and Cuddle on: Social Touch as a Stress Buffer', *Adaptive Human Behavior and Physiology*. Springer International Publishing, 2(4), pp. 344–362. doi: 10.1007/s40750-016-0052-x.
37. Beetz, A. *et al.* (2012) 'Psychosocial and psychophysiological effects of human-animal interactions: The possible role of oxytocin', *Frontiers in Psychology*. doi: 10.3389/fpsyg.2012.00234.
38. Black, D. S. and Slavich, G. M. (2016) 'Mindfulness meditation and the immune system: a systematic review of randomized controlled trials', *Annals of the New York Academy of Sciences*. Blackwell Publishing Inc., 1373(1), pp. 13–24. doi: 10.1111/nyas.12998.
39. Kang, D. H. *et al.* (2011) 'Dose effects of relaxation practice on immune responses in women newly diagnosed with breast cancer: An exploratory study', *Oncology Nursing Forum*, 38(3). doi: 10.1188/11.ONF.E240-E252.

40. Ulrich, R. S. (1984) 'View through a window may influence recovery from surgery', *Science*, 224(4647), pp. 420–421. doi: 10.1126/science.6143402.
41. Velarde, M. D., Fry, G. and Tveit, M. (2007) 'Health effects of viewing landscapes – Landscape types in environmental psychology', *Urban Forestry and Urban Greening*. Elsevier GmbH, 6(4), pp. 199–212. doi: 10.1016/j.ufug.2007.07.001.
42. Cohen, N., Moynihan, J. A. and Ader, R. (1994) 'Pavlovian Conditioning of the Immune System', *International Archives of Allergy and Immunology*, 105(2), pp. 101–106. doi: 10.1159/000236811.
43. Vits, S. *et al.* (2011) 'Behavioural conditioning as the mediator of placebo responses in the immune system', *Philosophical Transactions of the Royal Society B: Biological Sciences*, pp. 1799–1807. doi: 10.1098/rstb.2010.0392.
44. Zschucke, E. *et al.* (2015) 'The stress-buffering effect of acute exercise: Evidence for HPA axis negative feedback', *Psychoneuroendocrinology*. Elsevier Ltd, 51, pp. 414–425. doi: 10.1016/j.psyneuen.2014.10.019.
45. Lunt, H. C. *et al.* (2010) '"Cross-adaptation": Habituation to short repeated cold-water immersions affects the response to acute hypoxia in humans', *Journal of Physiology*, 588(18), pp. 3605–3613. doi: 10.1113/jphysiol.2010.193458.
46. Kröger, M. *et al.* (2015) 'Whole-body Cryotherapy's enhancement of acute recovery of running performance in well-trained athletes', *International Journal of Sports Physiology and Performance*. Human Kinetics Publishers Inc., 10(5), pp. 605–612. doi: 10.1123/ijspp.2014-0392.
47. Lubkowska, A. *et al.* (2011) 'The effect of prolonged whole-body cryostimulation treatment with different amounts of sessions on chosen pro-and anti-inflammatory cytokines levels in healthy men', *Scandinavian Journal of Clinical and Laboratory Investigation*, 71(5), pp. 419–425. doi: 10.3109/00365513.2011.580859.
48. Pournot, H. *et al.* (2011) 'Correction: Time-Course of Changes in Inflammatory Response after Whole-Body Cryotherapy Multi Exposures following Severe Exercise', *PLoS ONE*. Public

Library of Science (PLoS), 6(11). doi: 10.1371/annotation/0adb3312-7d2b-459c-97f7-a09cfecf5881.

49. Hirvonen, H. E. *et al.* (2006) 'Effectiveness of different cryotherapies on pain and disease activity in active rheumatoid arthritis. A randomised single blinded controlled trial', *Clinical and Experimental Rheumatology*, 24(3), pp. 295–301.
50. Šrámek, P. *et al.* (2000) 'Human physiological responses to immersion into water of different temperatures', *European Journal of Applied Physiology*, 81(5), pp. 436–442. doi: 10.1007/s004210050065.
51. Hu, X., Goldmuntz, E. A. and Brosnan, C. F. (1991) 'The effect of norepinephrine on endotoxin-mediated macrophage activation', *Journal of Neuroimmunology*, 31(1), pp. 35–42. doi: 10.1016/0165-5728(91)90084-K.
52. Ksiezopolska-Orłowska, K. *et al.* (2016) 'Complex rehabilitation and the clinical condition of working rheumatoid arthritis patients: Does cryotherapy always overtop traditional rehabilitation?', *Disability and Rehabilitation*. Taylor and Francis Ltd, 38(11), pp. 1034–1040. doi: 10.3109/09638288.2015.1060265.
53. Gizińska, M. *et al.* (2015) 'Effects of Whole-Body Cryotherapy in Comparison with Other Physical Modalities Used with Kinesitherapy in Rheumatoid Arthritis', *BioMed Research International*. Hindawi Publishing Corporation, 2015. doi: 10.1155/2015/409174.
54. Braun, K.-P. *et al.* (2009) 'Whole-body cryotherapy in patients with inflammatory rheumatic disease. A prospective study.', *Medizinische Klinik*, 104(3), pp. 192–6. doi: 10.1007/s00063-009-1031-9.
55. Buijze, G. A. *et al.* (2016) 'The effect of cold showering on health and work: A randomized controlled trial', *PLoS ONE*. Public Library of Science, 11(9). doi: 10.1371/journal.pone.0161749.
56. Bouzigon, R. *et al.* (2014) 'The use of whole-body cryostimulation to improve the quality of sleep in athletes during high level standard competitions', *British Journal of Sports Medicine*. BMJ, 48(7), pp. 572.1–572. doi: 10.1136/bjsports-2014-093494.33.

57. Lombardi, G., Ziemann, E. and Banfi, G. (2017) 'Whole-body cryotherapy in athletes: From therapy to stimulation. An updated review of the literature', *Frontiers in Physiology*. Frontiers Research Foundation. doi: 10.3389/fphys. 2017.00258.
58. Laukkanen, T. *et al.* (2015) 'Association between sauna bathing and fatal cardiovascular and all-cause mortality events', *JAMA Internal Medicine*. American Medical Association, 175(4), pp. 542–548. doi: 10.1001/jamainternmed.2014.8187.
59. Brunt, V. E. *et al.* (2016) 'Passive heat therapy improves endothelial function, arterial stiffness and blood pressure in sedentary humans', *Journal of Physiology*. Blackwell Publishing Ltd, 594(18), pp. 5329–5342. doi: 10.1113/JP272453.
60. Faulkner, S. H. *et al.* (2017) 'The effect of passive heating on heat shock protein 70 and interleukin-6: A possible treatment tool for metabolic diseases?', *Temperature*, 4(3), pp. 292–304. doi: 10.1080/23328940.2017.1288688.
61. Leicht, C. A. *et al.* (2017) 'Increasing heat storage by wearing extra clothing during upper body exercise up-regulates heat shock protein 70 but does not modify the cytokine response', *Journal of Sports Sciences*. Routledge, 35(17), pp. 1752–1758. doi: 10.1080/02640414.2016.1235795.
62. Hauet-Broere, F. *et al.* (2006) 'Heat shock proteins induce T cell regulation of chronic inflammation', *Annals of the Rheumatic Diseases*. doi: 10.1136/ard.2006.058495.
63. Singh, R. *et al.* (2010) 'Anti-Inflammatory Heat Shock Protein 70 Genes are Positively Associated with Human Survival', *Current Pharmaceutical Design*. Bentham Science Publishers Ltd., 16(7), pp. 796–801. doi: 10.2174/138161210790883499.
64. Masuda, A. *et al.* (2005) 'The effects of repeated thermal therapy for patients with chronic pain', *Psychotherapy and Psychosomatics*, 74(5), pp. 288–294. doi: 10.1159/000086319.
65. Selsby, J. T. *et al.* (2007) 'Intermittent hyperthermia enhances skeletal muscle regrowth and attenuates oxidative damage following reloading', *Journal of Applied Physiology*, 102(4), pp. 1702–1707. doi: 10.1152/japplphysiol.00722.2006.
66. Laukkanen, J. A. and Laukkanen, T. (2018) 'Sauna bathing and systemic inflammation', *European Journal of Epidemiology*.

Springer Netherlands, pp. 351–353. doi: 10.1007/s10654-017-0335-y.
67. Kunutsor, S. K., Laukkanen, T. and Laukkanen, J. A. (2017) 'Frequent sauna bathing may reduce the risk of pneumonia in middle-aged Caucasian men: The KIHD prospective cohort study', *Respiratory Medicine*. W.B. Saunders Ltd, 132, pp. 161–163. doi: 10.1016/j.rmed.2017.10.018.
68. Kukkonen-Harjula, K. and Kauppinen, K. (1988) 'How the sauna affects the endocrine system.', *Annals of Clinical Research*, 20(4), pp. 262–6. Available at: http://www.ncbi.nlm.nih.gov/pubmed/3218898 (Accessed: 1 December 2019).

Chapter 6

1. De Heredia, F. P., Gómez-Martínez, S. and Marcos, A. (2012) 'Chronic and degenerative diseases: Obesity, inflammation and the immune system', in *Proceedings of the Nutrition Society*. Cambridge University Press, pp. 332–338. doi: 10.1017/S0029665112000092.
2. McGreevy, K. R. *et al.* (2019) 'Intergenerational transmission of the positive effects of physical exercise on brain and cognition', *Proceedings of the National Academy of Sciences of the United States of America*. National Academy of Sciences, 116(20), pp. 10103–10112. doi: 10.1073/pnas.1816781116.
3. Grazioli, E. *et al.* (2017) 'Physical activity in the prevention of human diseases: Role of epigenetic modifications', *BMC Genomics*. BioMed Central Ltd. doi: 10.1186/s12864-017-4193-5.
4. Hespe, G. E. *et al.* (2016) 'Exercise training improves obesity-related lymphatic dysfunction', *Journal of Physiology*. Blackwell Publishing Ltd, 594(15), pp. 4267–4282. doi: 10.1113/JP271757.
5. Zaccardi, F. *et al.* (2019) 'Comparative Relevance of Physical Fitness and Adiposity on Life Expectancy: A UK Biobank Observational Study', *Mayo Clinic Proceedings*. Elsevier Ltd, 94(6), pp. 985–994. doi: 10.1016/j.mayocp.2018.10.029.
6. Barrett, B. *et al.* (2012) 'Meditation or exercise for preventing acute respiratory infection: A randomized controlled trial', *Annals of Family Medicine*. Annals of Family Medicine, Inc, 10(4), pp. 337–346. doi: 10.1370/afm.1376.

7. Nieman, D. C. *et al.* (2005) 'Immune response to a 30-minute walk', *Medicine and Science in Sports and Exercise*, 37(1), pp. 57–62. doi: 10.1249/01.MSS.0000149808.38194.21.
8. Pascoe, A. R., Fiatarone Singh, M. A. and Edwards, K. M. (2014) 'The effects of exercise on vaccination responses: A review of chronic and acute exercise interventions in humans', *Brain, Behavior, and Immunity*. Academic Press Inc., pp. 33–41. doi: 10.1016/j.bbi.2013.10.003.
9. Edwards, K. M. *et al.* (2006) 'Acute stress exposure prior to influenza vaccination enhances antibody response in women', *Brain, Behavior, and Immunity*, 20(2), pp. 159–168. doi: 10.1016/j.bbi.2005.07.001.
10. Edwards, K. M. *et al.* (2007) 'Eccentric exercise as an adjuvant to influenza vaccination in humans', *Brain, Behavior, and Immunity*, 21(2), pp. 209–217. doi: 10.1016/j.bbi.2006.04.158.
11. Pedersen, L. *et al.* (2016) 'Voluntary running suppresses tumor growth through epinephrine- and IL-6-dependent NK cell mobilization and redistribution', *Cell Metabolism*. Cell Press, 23(3), pp. 554–562. doi: 10.1016/j.cmet.2016.01.011.
12. Campbell, J. P. and Turner, J. E. (2018) 'Debunking the myth of exercise-induced immune suppression: Redefining the impact of exercise on immunological health across the lifespan', *Frontiers in Immunology*. Frontiers Media S.A. doi: 10.3389/fimmu.2018.00648.
13. Schafer, M. J. *et al.* (2016) 'Exercise prevents diet-induced cellular senescence in adipose tissue', *Diabetes*. American Diabetes Association Inc., 65(6), pp. 1606–1615. doi: 10.2337/db15-0291.
14. Cottam, M. A. *et al.* (2018) 'Links between Immunologic Memory and Metabolic Cycling', *The Journal of Immunology*. The American Association of Immunologists, 200(11), pp. 3681–3689. doi: 10.4049/jimmunol.1701713.
15. Gleeson, M., McFarlin, B. and Flynn, M. (2006) 'Exercise and toll-like receptors', *Exercise Immunology Review*, pp. 34–53.
16. Duggal, N. A. *et al.* (2018) 'Major features of immunesenescence, including reduced thymic output, are ameliorated by high levels of physical activity in adulthood', *Aging Cell*, 17(2), p. e12750. doi: 10.1111/acel.12750.

17. Bennett, J. A. and Winters-Stone, K. (2011) 'Motivating older adults to exercise: what works?', *Age and Ageing*, 40(2), pp. 148–149. doi: 10.1093/ageing/afq182.
18. Zhao, M. *et al.* (2019) 'Beneficial associations of low and large doses of leisure time physical activity with all-cause, cardiovascular disease and cancer mortality: A national cohort study of 88,140 US adults', *British Journal of Sports Medicine*. BMJ Publishing Group, 53(22), pp. 1405–1411. doi: 10.1136/bjsports-2018-099254.
19. Robinson, M. M. *et al.* (2017) 'Enhanced Protein Translation Underlies Improved Metabolic and Physical Adaptations to Different Exercise Training Modes in Young and Old Humans', *Cell Metabolism*. Cell Press, 25(3), pp. 581–592. doi: 10.1016/j.cmet.2017.02.009.
20. Rall, L. C. *et al.* (1996) 'Effects of progressive resistance training on immune response in aging and chronic inflammation', *Medicine and Science in Sports and Exercise*. Lippincott Williams and Wilkins, 28(11), pp. 1356–1365. doi: 10.1097/00005768-199611000-00003.
21. Piasecki, J. *et al.* (2019) 'Comparison of Muscle Function, Bone Mineral Density and Body Composition of Early Starting and Later Starting Older Masters Athletes', *Frontiers in Physiology*, 10. doi: 10.3389/fphys.2019.01050.
22. Guthold, R. *et al.* (2018) 'Worldwide trends in insufficient physical activity from 2001 to 2016: a pooled analysis of 358 population-based surveys with 1·9 million participants', *The Lancet Global Health*. Elsevier Ltd, 6(10), pp. e1077–e1086. doi: 10.1016/S2214-109X(18)30357-7.
23. Vairo, G. L. *et al.* (2009) 'Systematic Review of Efficacy for Manual Lymphatic Drainage Techniques in Sports Medicine and Rehabilitation: An Evidence-Based Practice Approach', *Journal of Manual & Manipulative Therapy*. Informa UK Limited, 17(3), pp. 80E–89E. doi: 10.1179/jmt.2009.17.3.80e.
24. Poppendieck, W. *et al.* (2016) 'Massage and Performance Recovery: A Meta-Analytical Review', *Sports Medicine*. Springer International Publishing, pp. 183–204. doi: 10.1007/s40279-015-0420-x.

25. Abolins, S. *et al.* (2018) 'The ecology of immune state in a wild mammal, Mus musculus domesticus', *PLOS Biology*. Edited by D. Schneider, 16(4), p. e2003538. doi: 10.1371/journal.pbio.2003538.
26. Fuss, J. *et al.* (2015) 'A runner's high depends on cannabinoid receptors in mice', *Proceedings of the National Academy of Sciences of the United States of America*. National Academy of Sciences, 112(42), pp. 13105–13108. doi: 10.1073/pnas.1514996112.
27. Dietrich, A. and McDaniel, W. F. (2004) 'Endocannabinoids and exercise', *British Journal of Sports Medicine*, pp. 536–541. doi: 10.1136/bjsm.2004.011718.
28. Raichlen, D. A. *et al.* (2012) 'Wired to run: Exercise-induced endocannabinoid signaling in humans and cursorial mammals with implications for the "runner's high"', *Journal of Experimental Biology*, 215(8), pp. 1331–1336. doi: 10.1242/jeb.063677.
29. Mørch, H. and Pedersen, B. K. (1995) 'βendorphin and the immune system – possible role in autoimmune diseases', *Autoimmunity*. Informa Healthcare, 21(3), pp. 161–171. doi: 10.3109/08916939509008013.
30. Schwarz, L. and Kindermann, W. (1992) 'Changes in β-Endorphin Levels in Response to Aerobic and Anaerobic Exercise', *Sports Medicine: An International Journal of Applied Medicine and Science in Sport and Exercise*, pp. 25–36. doi: 10.2165/00007256-199213010-00003.
31. Kraemer, W. J. *et al.* (1992) 'Acute hormonal responses in elite junior weightlifters', *International Journal of Sports Medicine*, 13(2), pp. 103–109. doi: 10.1055/s-2007-1021240.
32. Ekblom, B., Ekblom, Ö. and Malm, C. (2006) 'Infectious episodes before and after a marathon race', *Scandinavian Journal of Medicine and Science in Sports*, 16(4), pp. 287–293. doi: 10.1111/j.1600-0838.2005.00490.x.
33. Fahlman, M., Engels, H. and Hall, H. (2017) 'SIgA and Upper Respiratory Syndrome During a College Cross Country Season', *Sports Medicine International Open*. Georg Thieme Verlag KG, 1(06), pp. E188–E194. doi: 10.1055/s-0043-119090.

34. Cavaglieri, C. R. *et al.* (2011) 'Immune parameters, symptoms of upper respiratory tract infections, and training-load indicators in volleyball athletes', *International Journal of General Medicine*. Dove Medical Press Ltd., p. 837. doi: 10.2147/ijgm.s24402.
35. Parker, S., Brukner, P. and Rosier, M. (1996) 'Chronic fatigue syndrome and the athlete', *Sports Medicine, Training and Rehabilitation*. Taylor and Francis Ltd., 6(4), pp. 269–278. doi: 10.1080/15438629609512057.
36. Meeusen, R. *et al.* (2013) 'Prevention, diagnosis, and treatment of the overtraining syndrome: Joint consensus statement of the European College of Sport Science and the American College of Sports Medicine', *Medicine and Science in Sports and Exercise*, 45(1), pp. 186–205. doi: 10.1249/MSS.0b013e318279a10a.
37. Costa, R. J. S. *et al.* (2017) 'Systematic review: exercise-induced gastrointestinal syndrome-implications for health and intestinal disease', *Alimentary Pharmacology & Therapeutics*, 46(3), pp. 246–265. doi: 10.1111/apt.14157.
38. Spence, L. *et al.* (2007) 'Incidence, etiology, and symptomatology of upper respiratory illness in elite athletes', *Medicine and Science in Sports and Exercise*, 39(4), pp. 577–586. doi: 10.1249/mss.0b013e31802e851a.
39. Svendsen, I. S. *et al.* (2016) 'Training-related and competition-related risk factors for respiratory tract and gastrointestinal infections in elite cross-country skiers', *British Journal of Sports Medicine*. BMJ Publishing Group, 50(13), pp. 809–815. doi: 10.1136/bjsports-2015-095398.
40. Cavaglieri, C. R. *et al.* (2011) 'Immune parameters, symptoms of upper respiratory tract infections, and training-load indicators in volleyball athletes', *International Journal of General Medicine*. Dove Medical Press Ltd., p. 837. doi: 10.2147/ijgm.s24402.
41. Marqués-Jiménez, D. *et al.* (2016) 'Are compression garments effective for the recovery of exercise-induced muscle damage? A systematic review with meta-analysis', *Physiology and Behavior*. Elsevier Inc., pp. 133–148. doi: 10.1016/j.physbeh.2015.10.027.

42. Costello, J. T. *et al.* (2015) 'Whole-body cryotherapy (extreme cold air exposure) for preventing and treating muscle soreness after exercise in adults', *Cochrane Database of Systematic Reviews*. doi: 10.1002/14651858.CD010789.pub2.
43. Hohenauer, E. *et al.* (2015) 'The effect of post-exercise cryotherapy on recovery characteristics: A systematic review and meta-analysis', *PLoS ONE*. Public Library of Science, 10(9). doi: 10.1371/journal.pone.0139028.
44. Gunzer, W., Konrad, M. and Pail, E. (2012) 'Exercise-induced immunodepression in endurance athletes and nutritional intervention with carbohydrate, protein and fat – what is possible, what is not?', *Nutrients*. MDPI AG, pp. 1187–1212. doi: 10.3390/nu4091187.
45. Weidner, T. G. *et al.* (1998) 'The effect of exercise training on the severity and duration of a viral upper respiratory illness', *Medicine and Science in Sports and Exercise*. American College of Sports Medicine, 30(11), pp. 1578–1583. doi: 10.1097/00005768-199811000-00004.
46. Arrieta, M. C., Bistritz, L. and Meddings, J. B. (2006) 'Alterations in intestinal permeability', *Gut*, pp. 1512–1520. doi: 10.1136/gut.2005.085373.
47. Sharif, K. *et al.* (2017) 'Physical activity and autoimmune diseases: Get moving and manage the disease'. doi: 10.1016/j.autrev.2017.11.010.
48. Iversen, M. D. *et al.* (2017) 'Physical Activity and Correlates of Physical Activity Participation Over Three Years in Adults with Rheumatoid Arthritis', *Arthritis Care and Research*. John Wiley and Sons Inc., 69(10), pp. 1535–1545. doi: 10.1002/acr.23156.
49. O'Neill, H. M. (2013) 'AMPK and exercise: Glucose uptake and insulin sensitivity', *Diabetes and Metabolism Journal*, pp. 1–21. doi: 10.4093/dmj.2013.37.1.1.
50. Johnson, J. L. *et al.* (2007) 'Exercise Training Amount and Intensity Effects on Metabolic Syndrome (from Studies of a Targeted Risk Reduction Intervention through Defined Exercise)', *American Journal of Cardiology*, 100(12), pp. 1759–1766. doi: 10.1016/j.amjcard.2007.07.027.

Chapter 7

1. Hemilä, H. (1997) 'Vitamin C supplementation and the common cold – Was Linus Pauling right or wrong?', *International Journal for Vitamin and Nutrition Research*, 67(5), pp. 329–335.
2. Hemilä, H. and Chalker, E. (2013) 'Vitamin C for preventing and treating the common cold', *Cochrane Database of Systematic Reviews*. John Wiley and Sons Ltd. doi: 10.1002/14651858.CD000980.pub4.
3. Mocchegiani, E. (2007) 'Zinc and ageing: Third Zincage conference', *Immunity and Ageing*, 4. doi: 10.1186/1742-4933-4-5.
4. Singh, M. and Das, R. R. (2011) 'Zinc for the common cold', in Singh, M. (ed.) *Cochrane Database of Systematic Reviews*. Chichester, UK: John Wiley & Sons, Ltd. doi: 10.1002/14651858.CD001364.pub3.
5. Public Health England (2016) 'National Diet and Nutrition Survey – GOV.UK', *Public Health England*, 3, pp. 1–79. Available at: https://www.gov.uk/government/collections/national-diet-and-nutrition-survey (Accessed: 16 December 2019).
6. Maggini, S., Pierre, A. and Calder, P. (2018) 'Immune Function and Micronutrient Requirements Change over the Life Course', *Nutrients*, 10(10), p. 1531. doi: 10.3390/nu10101531.
7. Arola-Arnal, A. *et al.* (2019) 'Chrononutrition and Polyphenols: Roles and Diseases', *Nutrients*, 11(11), p. 2602. doi: 10.3390/nu11112602.
8. Barański, M. *et al.* (2014) 'Higher antioxidant and lower cadmium concentrations and lower incidence of pesticide residues in organically grown crops: A systematic literature review and meta-analyses', *British Journal of Nutrition*. Cambridge University Press, pp. 794–811. doi: 10.1017/S0007114514001366.
9. Ristow, M. *et al.* (2009) 'Antioxidants prevent health-promoting effects of physical exercise in humans', *Proceedings of the National Academy of Sciences of the United States of America*, 106(21), pp. 8665–8670. doi: 10.1073/pnas.0903485106.

10. Peternelj, T. T. and Coombes, J. S. (2011) 'Antioxidant supplementation during exercise training: Beneficial or detrimental?', *Sports Medicine*, pp. 1043–1069. doi: 10.2165/11594400-000000000-00000.
11. Wu, G. (2013) 'Arginine and immune function', in *Diet, Immunity and Inflammation*. Elsevier Ltd., pp. 523–543. doi: 10.1533/9780857095749.3.523.
12. Cruzat, V. *et al.* (2018) 'Glutamine: Metabolism and immune function, supplementation and clinical translation', *Nutrients*. MDPI AG. doi: 10.3390/nu10111564.
13. Brigham, E. P. *et al.* (2019) 'Omega-3 and Omega-6 Intake Modifies Asthma Severity and Response to Indoor Air Pollution in Children', *American Journal of Respiratory and Critical Care Medicine*, 199(12), pp. 1478–1486. doi: 10.1164/rccm.201808-1474OC.
14. Sun, Q., Li, J. and Gao, F. (2014) 'New insights into insulin: The anti-inflammatory effect and its clinical relevance', *World Journal of Diabetes*. Baishideng Publishing Group Inc., 5(2), p. 89. doi: 10.4239/wjd.v5.i2.89.
15. Aeberli, I. *et al.* (2011) 'Low to moderate sugar-sweetened beverage consumption impairs glucose and lipid metabolism and promotes inflammation in healthy young men: A randomized controlled trial', *American Journal of Clinical Nutrition*, 94(2), pp. 479–485. doi: 10.3945/ajcn.111.013540.
16. Duncan, S. H. *et al.* (2007) 'Reduced Dietary Intake of Carbohydrates by Obese Subjects Results in Decreased Concentrations of Butyrate and Butyrate-Producing Bacteria in Feces', *Applied and Environmental Microbiology*, 73(4), pp. 1073–1078. doi: 10.1128/AEM.02340-06.
17. Catassi, C. *et al.* (2013) 'Non-celiac gluten sensitivity: The new frontier of gluten related disorders', *Nutrients*. MDPI AG, pp. 3839–3853. doi: 10.3390/nu5103839.
18. Sapone, A. *et al.* (2012) 'Spectrum of gluten-related disorders: consensus on new nomenclature and classification', *BMC Medicine*, 10(1), p. 13. doi: 10.1186/1741-7015-10-13.
19. Sapone, A. *et al.* (2011) 'Divergence of gut permeability and mucosal immune gene expression in two gluten-associated

conditions: celiac disease and gluten sensitivity', *BMC Medicine*, 9(1), p. 23. doi: 10.1186/1741-7015-9-23.

20. Dixit, V. D. *et al.* (2011) 'Controlled meal frequency without caloric restriction alters peripheral blood mononuclear cell cytokine production', *Journal of Inflammation*, 8. doi: 10.1186/1476-9255-8-6.
21. Willebrand, R. and Kleinewietfeld, M. (2018) 'The role of salt for immune cell function and disease', *Immunology*. Blackwell Publishing Ltd, pp. 346–353. doi: 10.1111/imm.12915.
22. Yosef, N. *et al.* (2013) 'Dynamic regulatory network controlling TH 17 cell differentiation', *Nature*, 496(7446), pp. 461–468. doi: 10.1038/nature11981.
23. Wu, C. *et al.* (2013) 'Induction of pathogenic TH 17 cells by inducible salt-sensing kinase SGK1', *Nature*, 496(7446), pp. 513–517. doi: 10.1038/nature11984.
24. Sundstrom, B., Johansson, I. and Rantapaa-Dahlqvist, S. (2015) 'Interaction between dietary sodium and smoking increases the risk for rheumatoid arthritis: results from a nested case-control study', *Rheumatology*, 54(3), pp. 487–493. doi: 10.1093/rheumatology/keu330.
25. Gill, S. and Panda, S. (2015) 'A Smartphone App Reveals Erratic Diurnal Eating Patterns in Humans that Can Be Modulated for Health Benefits', *Cell Metabolism*. Cell Press, 22(5), pp. 789–798. doi: 10.1016/j.cmet.2015.09.005.
26. Casas, R., Sacanella, E. and Estruch, R. (2014) 'The Immune Protective Effect of the Mediterranean Diet against Chronic Low-grade Inflammatory Diseases', *Endocrine, Metabolic & Immune Disorders-Drug Targets*. Bentham Science Publishers Ltd., 14(4), pp. 245–254. doi: 10.2174/1871530314666140922153350.
27. Sureda, A. *et al.* (2018) 'Adherence to the Mediterranean diet and inflammatory markers', *Nutrients*. MDPI AG, 10(1). doi: 10.3390/nu10010062.
28. Mutlu, E. A. *et al.* (2012) 'Colonic microbiome is altered in alcoholism', *American Journal of Physiology – Gastrointestinal and Liver Physiology*, 302(9). doi: 10.1152/ajpgi.00380.2011.
29. Queipo-Ortuño, M. I. *et al.* (2012) 'Influence of red wine polyphenols and ethanol on the gut microbiota ecology and

biochemical biomarkers', *American Journal of Clinical Nutrition*, 95(6), pp. 1323–1334. doi: 10.3945/ajcn.111.027847.

30. Zopf, Y. *et al.* (2009) 'Differenzialdiagnose von nahrungsmittelunverträglichkeiten', *Deutsches Arzteblatt*, 106(21), pp. 359–370. doi: 10.3238/arztebl.2009.0359.
31. Venter, C. *et al.* (2008) 'Prevalence and cumulative incidence of food hypersensitivity in the first 3 years of life', *Allergy: European Journal of Allergy and Clinical Immunology*, 63(3), pp. 354–359. doi: 10.1111/j.1398-9995.2007.01570.x.
32. Isolauri, E. *et al.* (1998) 'Elimination diet in cow's milk allergy: Risk for impaired growth in young children', *Journal of Pediatrics*. Mosby Inc., 132(6), pp. 1004–1009. doi: 10.1016/S0022-3476(98)70399-3.
33. Ozdemir, O. *et al.* (2009) 'Food intolerances and eosinophilic esophagitis in childhood', *Digestive Diseases and Sciences*, pp. 8–14. doi: 10.1007/s10620-008-0331-x.
34. Savaiano, D. A. *et al.* (2013) 'Improving lactose digestion and symptoms of lactose intolerance with a novel galacto-oligosaccharide (RP-G28): A randomized, double-blind clinical trial', *Nutrition Journal*. BioMed Central Ltd., 12(1). doi: 10.1186/1475-2891-12-160.
35. Fitzgerald, M. and Frankum, B. (2017) 'Food avoidance and restriction in adults: a cross-sectional pilot study comparing patients from an immunology clinic to a general practice', *Journal of Eating Disorders*, 5(1), p. 30. doi: 10.1186/s40337-017-0160-4.
36. First, M. B. (2013) *DSM-5® Handbook of Differential Diagnosis*. American Psychiatric Publishing. doi: 10.1176/appi.books.9781585629992.
37. Hammond, C. and Lieberman, J. A. (2018) 'Unproven Diagnostic Tests for Food Allergy', *Immunology and Allergy Clinics of North America*. W.B. Saunders, pp. 153–163. doi: 10.1016/j.iac.2017.09.011.
38. Lavine, E. (2012) 'Primer: Blood testing for sensitivity, allergy or intolerance to food', *CMAJ*. Canadian Medical Association, 184(6), pp. 666–668. doi: 10.1503/cmaj.110026.
39. Mullin, G. E. *et al.* (2010) 'Testing for food reactions: The good, the bad, and the ugly', *Nutrition in Clinical Practice*,

pp. 192–198. doi: 10.1177/0884533610362696.
40. Stapel, S. O. *et al.* (2008) 'Testing for IgG4 against foods is not recommended as a diagnostic tool: EAACI Task Force Report*', *Allergy*, 63(7), pp. 793–796. doi: 10.1111/j.1398-9995.2008.01705.x.
41. Gibson, P. R. *et al.* (2007) 'Review article: Fructose malabsorption and the bigger picture', *Alimentary Pharmacology and Therapeutics*, pp. 349–363. doi: 10.1111/j.1365-2036.2006.03186.x.
42. Michel, C. *et al.* (2014) 'A taste of Kandinsky: assessing the influence of the artistic visual presentation of food on the dining experience', *Flavour*, 3(1), p. 7. doi: 10.1186/2044-7248-3-7.

Further Reading and Resources

- Parkrun https://www.parkrun.org.uk
- National Health Service (NHS) https://www.nhs.uk
- World Health Organisation (WHO) https://www.who.int
- The National Institute for Health and Care Excellence (NICE) https://www.nice.org.uk
- Diagnostic and Statistical Manual of Mental Disorders-5 (DSM-5) https://www.psychiatry.org/psychiatrists/practice/dsm
- British Society for Immunology – a useful resource for the public seeking information on any immune related issues.
- MIND mental health charity www.mind.org.uk
- National Trust have a Beginners guide to Forest Bathing https://www.nationaltrust.org.uk/lists/a-beginners-guide-to-forest-bathing

Acknowledgements

To Luca and Isabella, my beautiful children who have given me the deepest sense of purpose and make all the hard times worth it.

I am forever grateful to my husband for his tireless patience in supporting me and reassuring me (and sometimes eye rolling me) when I take on new projects (like this one) on top of my already busy schedule. You have been integral not only to helping me manage my work-life juggle but also for believing in me and giving me the confidence to chase after a dream.

A special thanks to my parents for everything they have given me. Particularly for your support and encouragement of my never ending curiosity with the human body since early childhood. You have given me the grit to never give up and made me see that anything can be possible when you try.

Thanks also to my former supervisor and mentor Professor Clare Lloyd at Imperial College London. Plus all my 'Imperial Girls' for all the laughs over the years which helped lighten those long hard days in the lab. Thank you to all the many wonderful colleagues, scientists and health professionals whom I have had the privilege to work with and learn from in my career. And to the many great and inspirational immunologists whose work has deeply inspired my love of the subject over the last 20 years.

Massive thanks to my wonderful literary agent Carly. Your straight-talking advice and encouragement has been invaluable in making this dream come true. Carolyn and the brilliant publishing team at HarperCollins for their patience, guidance and support. Huge thanks to Rupy, Hazel and Rhiannon for letting me excitedly talk about immunity on their social platforms. And to Natasha at Neon Rocks for her wonderful and gentle approach to wellbeing PR.

I am especially grateful to all my dear friends for helping me in more ways than I thought possible and making life infinitely better. Importantly, I am forever indebted to all those who believed in me when I struggled to believe in myself. You are too numerous to name, but I will make sure you know who you are.

Index